COMPLETE GUIDE TO ACTIVE FILTER DESIGN, OP AMPS, AND PASSIVE COMPONENTS

COMPLETE GUIDE TO ACTIVE FILTER DESIGN, OP AMPS, AND PASSIVE COMPONENTS

Z. H. MEIKSIN

PRENTICE HALL
Englewood Cliffs, New Jersey 07632

Library of Congress Cataloging-in-Publication Data

Meiksin, Z. H.
  Complete guide to active filter design, op amps, and passive
components / Z.H. Meiksin.
    p. cm.
  Includes index.
  ISBN 0-13-159971-2
  1. Electric filters, Active—Deisgn and construction.
2. Operational amplifiers.   I. Title.
TK7872.F5M45   1990
621.381'5324—dc20   89-8646
                                                          CIP

Editorial/production supervision and
  interior design: Lorraine Antine and Kerry Reardon
Cover design: 20/20 Services
Manufacturing buyer: Robert Anderson

© 1990 by Prentice-Hall, Inc.
A Division of Simon & Schuster
Englewood Cliffs, New Jersey 07632

The publisher offers discounts on this book when ordered
in bulk quantities. For more information, write:

  Special Sales/College Marketing
  College Technical and Reference Division
  Prentice Hall
  Englewood Cliffs, New Jersey 07632

10 9 8 7 6 5 4 3 2 1

ISBN   0-13-159971-2

PRENTICE HALL INTERNATIONAL (UK) LIMITED, LONDON
PRENTICE HALL OF AUSTRALIA PTY. LIMITED, SYDNEY
PRENTICE HALL CANADA INC., TORONTO
PRENTICE HALL HISPANOAMERICANA, S.A., MEXICO
PRENTICE HALL OF INDIA PRIVATE LIMITED, NEW DELHI
PRENTICE HALL OF JAPAN, INC., TOKYO
SIMON & SCHUSTER ASIA PTE. LTD. SINGAPORE
EDITORA PRENTICE HALL DO BRASIL, LTDA., RIO DE JANEIRO

DEDICATION
To my granddaughter Rebecca

Contents

Preface

Essentially all electronic signal-processing systems incorporate filters of one sort or another to separate a selected group of voltages from the remainder, or to improve signal-to-noise ratio. Readily available, high-quality operational amplifiers (also called op amps) of small size and low cost make the use of active filters very attractive. The theory of active filters is complex. But the practical design of such filters, if presented properly, is simple and straightforward. This book provides such a practical approach to filter design.

Active filters are built with operational amplifiers, resistors, and capacitors. In the real world these components are not ideal. To achieve performance within specifications, the real properties of the components must be taken into consideration in the design. To provide you with the information needed to design active filters properly, this handbook is organized in four parts and two appendices.

Part I of this practical guide shows you how to select and design the right filter. If you are designing an amplifying system for an audio frequency signal above 500 Hz, for example, you can filter out 60-Hz line-frequency noise voltage from the power supply by incorporating in your system a simple first-order high-pass filter. There is no need to spend time and money for a higher-order filter. If, on the other hand, you build a system in which you want to process voltages with frequencies between 50 kHz and 75 kHz and you want to reduce the shot noise present in all electronic components, you need a filter with sharper cutoff characteristics. You can incorporate in the system a third-order band-pass filter and you can even, with the formulas given in this book, compute the exact attenuation for all frequencies.

Reading Chapter 1 first will enable you to capitalize more quickly on the useful guidelines and specific data provided in this book. This chapter defines *filter order*, *filter classes*, and the *normalized filter*, as well as the basic concepts of low-pass, high-pass, and band-pass filters. You decide what type of filter you need and you go directly to the chapter and section that directs you step-by-step to the desired circuit configuration and, by means of simple equations or tables, to the determination of the component values.

First-order filters, given in Chapter 2, are the easiest to design. The selection of components for the desired cutoff frequencies are obtained by the use of simple formulas. To design second-order filters, given in Chapter 3, as well as to design higher-order filters, the *normalized filter* concept is used. This approach employs three distinct, clearly defined steps, which are explained with examples. Third-order filters are presented in Chapter 4. These are the "work horses" of filter design. They are simple, economical, and effective for a wide range of applications. For sharper cutoff characteristics, fourth-order filters, given in Chapter 5, are used. For still sharper cutoff characteristics, higher-order filters must be used.

How to design fifth- through tenth-order filters is explained in Chapter 6. Using a multitude of tables would defocus the essential aspects of this design guide. Therefore, a single table is given for all classes and higher order filters, with special equations for all filter stages. It is explained clearly with the aid of examples how to use these equations to arrive at the filter component values in a most straightforward manner. The chapter concludes with the high-Q state variable filter.

Chapter 7 shows how amplifier and passive-component performance affects filter performance and how to avoid pitfalls.

Part II deals with operational amplifiers. It shows you how to interpret manufacturers' data sheets, how to evaluate operational amplifier performance, and how to incorporate real operational amplifier characteristics into practical circuit design. The extensive material covered in the five chapters of this part is useful not only for filter design, but for electronic circuit design in general. Dual-power-supply as well as single-power-supply and special low-voltage amplifiers—popular in battery-operated systems—are included. Chapter 12 gives complete design examples, which take into consideration the practical limitations and imperfections of operational amplifiers.

Part III introduces and explains passive components. Although active filters use only resistors and capacitors, this part also presents details of inductors, transformers, and diodes, which makes this chapter extremely important for electronic circuit design in general. These components are made of a large number of different materials using a variety of technologies. Materials are complicated and their properties are influenced by fabrication processes that in turn influence component properties. The properties of all these passive components are explained, guiding you to the correct choice of components in your designs. You are shown how to derate power ratings of resistors with increase in temperature and how the resistance of carbon resistors depends on frequency. Capacitor parameters are defined, and curves show the dependency of capacitance and insulation resistance on variables such as frequency and temperature. Also given is the dependency of the dissipation factor on the applied DC voltage. You will even find how to measure values and characteristics of passive components using techniques ranging from simple ohmmeter measurements to sophisticated methods using more complex, but readily available, instruments. Also included in the seven chapters of this part of the handbook is the subject of component identification. A variety of color, symbol, and letter codes are used by manufacturers to identify component values, tolerances, and temperature coefficients. With the help of the information given, you will be able to identify virtually any commercial component.

The material in this design guide is carefully pulled together in Part IV. Design examples that lead you from the design specifications through the final product are given. You are shown all the design steps and how to choose particular op amps to ensure that the specifications are being met. You are shown how to select the necessary resistors and capacitors, including values, voltage ratings, power ratings, and the particular material from which the components are made, to meet special requirements of the design. Effects of temperature and component tolerances are illustrated.

The utility of this well-planned book is further enhanced by the inclusion of two appendices. Appendix A includes a large variety of manufacturer data sheets for op amps selected carefully to give you representative data of the most important op amp types. Appendix B compiles symbols and codes to make quick passive component identification possible.

The detailed explanations of the material, the data sheets, design procedures, formulas, diagrams and tables make this practical handbook an invaluable design aid in your present and future work.

ACKNOWLEDGMENT
The contributions and suggestions
made by Philip C. Thackray are
gratefully acknowledged.

COMPLETE GUIDE TO ACTIVE FILTER DESIGN, OP AMPS, AND PASSIVE COMPONENTS

Part I HOW TO DESIGN ACTIVE FILTERS

1

Understanding Filter Terminology

1.1 INTRODUCTION

The successful design of filters requires the designer to clearly formulate in the mind, and then on paper, the intended use of the filter. A prescribed set of steps is then followed to complete the design. This process is enhanced by thinking in terms of concepts that are defined and expounded on in this chapter. After the filter is designed and assembled, fine tuning is required for demanding applications. This is also covered in this chapter.

1.2 LOW-PASS, HIGH-PASS, AND BAND-PASS FILTERS

The ideal transfer characteristics of *low-pass, high-pass,* and *band-pass* filters are shown in Figure 1-1. Ideally, the signal is not attenuated at all in the passband of a filter, and is attenuated completely in the stop band. Also, ideally, the phase shift of the signal is linear with frequency in the passband. The phase shift in the stop band has no significance, since the signal is not transmitted in this band. Under these ideal conditions, the signal is transmitted without any distortion in the passband frequency range, and it is not transmitted at all outside this frequency range. In practice, the ideal filter can only be approached to various degrees, depending upon the complexity of the filter.

1.3 FILTER ORDER

The simplest filter contains *one* capacitor for each cutoff frequency. This type is called a *first-order* filter. Thus, a first-order low-pass filter contains one capacitor; a first-order high-pass filter contains one capacitor; and a first-order band-pass filter contains two

1

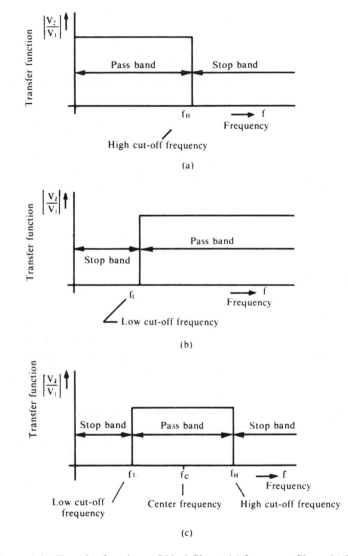

Figure 1-1 Transfer functions of ideal filters. (a) Low-pass filter. (b) High-pass filter. (c) Band-pass filter.

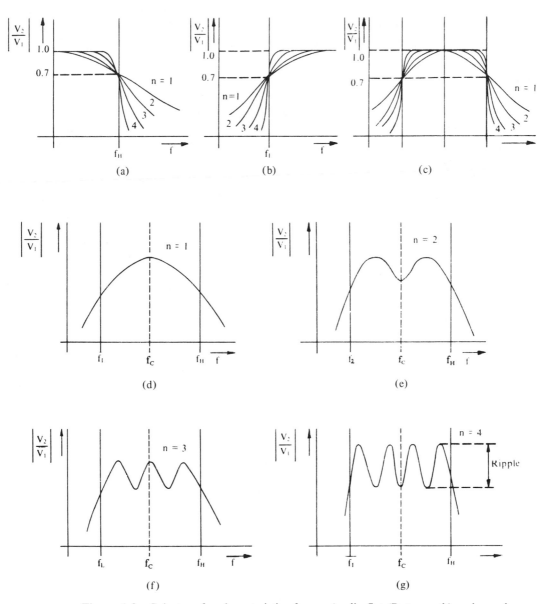

Figure 1-2 Gain transfer characteristics for *maximally flat* (Butterworth) and *equal-ripple* (Chebyshev) filters. (Not to scale.) (a) Low-pass maximally flat. (b) High-pass maximally flat. (c) Band-pass maximally flat. (d) Band-pass equal-ripple, $n = 1$. (e) Band-pass equal-ripple, $n = 2$. (f) Band-pass equal-ripple, $n = 3$. (g) Band-pass equal-ripple, $n = 4$.

capacitors. The next degree of complexity is obtained in the *second-order* filter, then the *third-order, fourth-order,* etc. The higher the order of the filter, the more nearly does the filtering characteristic of the filter approach that of the ideal filter.

1.4 FILTER CLASSES

A filter of *any given order* can be made to approximate the ideal filter in different ways, depending on the values given to the filter components. The type of approximation chosen determines the filter *class*. Two useful classes are the *maximally flat* (also known as Butterworth) and the *equal-ripple* (also known as Chebyshev) classes. The difference between the two classes is illustrated in Figure 1-2. Compare, for example, the gain-frequency characteristic of the second-order maximally flat band-pass filter given in Figure 1-2(c), ($n = 2$), with that of the second-order, equal-ripple, band-pass filter given in Figure 1-2(e). The maximally flat filter approximates the ideal filter better in the passband, but attenuates signals in the stop band less than the equal-ripple filter. In Figure 1-2 only band-pass diagrams for equal-ripple filters are shown. Low-pass and high-pass filter characteristics can be deduced by considering the appropriate portion of the corresponding band-pass filter characteristic. The frequency response curve of a fourth-order, low-pass maximally flat filter is compared with the response curve of a corresponding equal-ripple filter in Figure 1-3.

The particular order and class chosen for any given design depends on the particular system specifications. The design is usually carried out in terms of steady-state frequency-response specifications. However, the same filter components that determine the frequency response also determine the transient response. The mathematical relationships between the frequency response and the transient response are quite complex, and some experimentation in practical design is necessary. The more ''squared-off'' the frequency response is, the more overshoot is encountered during a transient period.

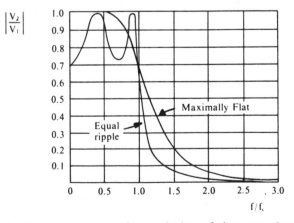

Figure 1-3 Response curves of magnitudes of low-pass fourth-order *maximally flat* (Butterworth) and *equal-ripple* (Chebyshev) filters.

1.5 NORMALIZED FILTERS

Filter design is simplified, particularly when designing higher-order filters, if the "normalized filter" concept is used. One first uses a simple set of design equations or tables to obtain the normalized filter. This filter has a cutoff frequency of 1 rad/s (or $1/2\pi = 0.159$ Hz) and usually resistors of 1 Ω and/or capacitors of 1 F. Two additional design steps result in the desired practical filter: first, a change of component values (resistors or capacitors) is made to obtain the *desired* cutoff frequency, and finally, a change in values of resistors and capacitors is made in a prescribed manner to obtain realistic orders of magnitude of component values.

1.6 FINE TUNING

The values resulting from the computations are given to accuracies considerably beyond the tolerances of practical components. In practice, one uses the closest commercially available components and checks to see whether the filter characteristics are still acceptable and, if necessary, adds either resistor trimmers or capacitor trimmers to *tune* the filter to the exact desired cutoff frequencies. Remember also that capacitor tolerances are on the order of $+80\% -20\%$, $\pm 20\%$, $\pm 10\%$, or $\pm 5\%$, and resistor tolerances $\pm 10\%$ or $\pm 5\%$, which would necessitate trimmers anyway. Resistors with $\pm 1\%$ tolerance are available at a higher cost. Don't overlook thick-film and thin-film resistors and capacitors if your product involves a large number of filters. Film components can be trimmed to order during the fabrication process, which can result in savings.

2
Designing First-Order Filters

2.1 INTRODUCTION

First-order filters are the simplest and least-expensive filters. They are suitable for nondemanding applications. In this chapter we show you how to design low-pass, high-pass, and band-pass first-order filters.

2.2 LOW-PASS FILTERS

The circuit diagram of a first-order low-pass filter is shown in Figure 2-1. The 3-dB cutoff frequency is given by

$$f_H = \frac{1}{2\pi R_2 C_2}$$

The DC gain is R_2/R_1, and the rolloff in the stop band is -20 dB/dec (-6 dB/oct).

Example

For a low-pass amplifier with $f_H = 1,000$ Hz and DC gain $= 10$,

$$\frac{1}{2\pi R_2 C_2} = 1,000$$

$$\frac{R_2}{R_1} = 10$$

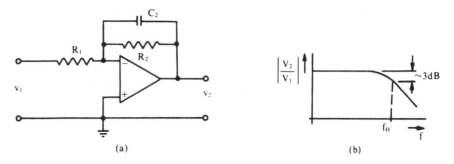

Figure 2-1 First-order low-pass filter. (a) Circuit diagram. (b) Frequency characteristic.

We have two equations with three unknowns. We can therefore choose arbitrarily any one of the three components. Let C_2 = 0.01 µF. From the equations above we obtain R_2 = 15.9 kΩ, and R_1 = 1.59 kΩ.

2.3 HIGH-PASS FILTERS

The circuit diagram of a first-order high-pass filter is given in Figure 2-2. The 3-dB cutoff frequency is given by

$$f_L = \frac{1}{2\pi R_1 C_1}$$

and the high-frequency gain beyond the cutoff frequency is R_2/R_1. The rolloff below the cutoff frequency is 20 dB/dec (6 dB/oct).

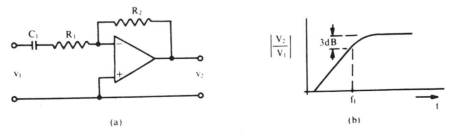

Figure 2-2 First-order high-pass filter. (a) Circuit diagram. (b) Frequency characteristic.

Example

For a high-pass amplifier with $f_L = 20$ kHz and high-frequency gain of 10,

$$\frac{1}{2\pi R_1 C_1} = 20,000$$

$$\frac{R_2}{R_1} = 10$$

Let $C_1 = 10,000$ pF. From the equations above, $R_1 = 796\ \Omega$ and $R_2 = 7.96$ kΩ.

2.4 BAND-PASS FILTERS

The circuit diagram of a first-order band-pass filter is shown in Figure 2-3. The low and high cutoff frequencies are, respectively,

$$f_L = \frac{1}{2\pi R_2 C_2} \quad \text{and} \quad f_H = \frac{1}{2\pi R_1 C_1}$$

The mid-frequency gain is R_2/R_1. There are two capacitors in this circuit, one associated with the low cutoff frequency, and one with the high cutoff frequency. The slopes of the transfer curve below f_L and above f_H are 20 dB/dec (6 dB/oct) and -20 dB/dec (6 dB/oct), respectively.

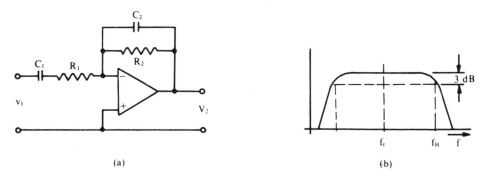

(a)   (b)

Figure 2-3 Simple band-pass filter. (a) Circuit diagram. (b) Frequency characteristic.

Example

It is desired to design a filter with f_L = 2 kHz, f_H = 5 kHz and mid-frequency gain = 5. The design equations are

$$\frac{1}{2\pi R_2 C_2} = 2{,}000$$

$$\frac{1}{2\pi R_1 C_1} = 5{,}000$$

$$\frac{R_2}{R_1} = 5$$

We have three equations with four unknown parameters. We can, therefore, choose any one of the four arbitrarily. Let C_2 = 0.01 μF, then R_2 = 7.958 kΩ, R_1 = 1.592 kΩ, and C_1 = 0.02 μF.

3

Designing Second-Order Filters

3.1 INTRODUCTION

Second-order filters require more components than first-order filters, but they possess sharper rolloff characteristics. In the stop band containing the frequency range in which signals are to be attenuated, the amplitudes of signals decrease faster with frequency than for first-order filters. The frequency responses in the passband and in the stop band can be shaped by the choice of the filter class: maximally flat (Butterworth), or equal-ripple (Chebyshev) with a specified degree of ripple. There are low-gain amplifier filters and high-gain amplifier filters. These two filter types are explained and design procedures with examples are given for both. Similarities between the two types and differences between them are highlighted, giving guidelines for their respective applications.

3.2 LOW-GAIN AMPLIFIER FILTERS

The term *low-gain amplifier* refers to an operational amplifier with an external feedback resistor and an external input resistor connected to the noninverting input of the amplifier. The gain of the amplifier is unity plus the ratio of the feedback resistor to the input resistor. The complete unit is now considered as a low-gain amplifier element of the filter which includes additional elements (resistors and capacitors) peripheral to the low-gain amplifier. This filter type allows us to use identical resistors throughout the circuit, which makes matching and temperature tracking better and, in large quantities, results in lower cost.

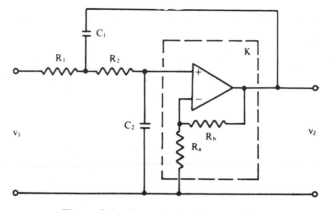

Figure 3-1   Second-order low-pass filter.

3.2.1 Low-Pass Filters

A second-order low-pass filter is shown in Figure 3-1. The gain of the amplifier is $K = 1 + R_b/R_a$.

The transfer function of the filter is determined by the value of the amplifier gain and by the values of the resistors and capacitors. The same transfer function can be realized with different combinations of these values. Three choices are particularly useful: (a) $K = 2$, which results in $R_a = R_b$, which in turn gives good matching and tracking; (b) $K = 1$, which results in a voltage-follower amplifier and eliminates the need for the two resistors R_a and R_b—R_a is replaced with an open circuit and R_b with a short circuit; and (c) $R_1 = R_2$ and $C_1 = C_2$, which again results in good matching and tracking. Further, whenever equal components are used, there will be a lower cost when a large number of filters are built. The filter design is first carried out for the cutoff frequency $\omega_H = 1$ rad/s. The component values are then changed in two steps: the first step gives the correct cutoff frequency, and the second step gives reasonable values of resistors and capacitors. Table 3-1 lists gain and component values for the three choices of $\omega_H = 1$ for the maximally flat filter and for two equal ripple filters, one with 0.5-dB ripple, and the other with 1-dB ripple.

Example

Design a low-pass, zero-ripple filter with a cutoff frequency of 1,500 Hz and a DC gain of 2.

The normalized filter is a maximally flat filter with the circuit configuration of Figure 3-1 and the values $R_1 = R_2 = 1.00000\ \Omega$, $C_1 = 0.87403$ F, and $C_2 = 1.1442$ F, as obtained from Table 3-1. To obtain a cutoff frequency of $\omega_H = 2\pi 1,500$, we divide all capacitors by $2\pi 1,500$ giving

$$C_1 = 0.87403/2\pi 1,500 = 92.7375\ \mu\text{F}$$
$$C_2 = 1.14412/2\pi 1,500 = 121.3949\ \mu\text{F}$$

TABLE 3-1 Component and Gain Values for the Low-Pass Filter of Figure 3-1 for $\omega_H = 1$ (R in ohms; C in farads)

Filter class	R_1	R_2	C_1	C_2	K
Maximally flat	1.00000	1.00000	0.87403	1.14412	2.00000
(Butterworth)	1.00000	1.00000	1.41421	0.70711	1.00000
3.01 dB at ω_H	1.00000	1.00000	1.00000	1.00000	1.58578
Equal ripple	1.00000	1.00000	0.77088	0.855557	2.00000
(Chebyshev)	1.00000	1.00000	1.40259	0.47013	1.00000
0.5-dB ripple	0.81220	0.81220	1.00000	1.00000	1.84213
Equal ripple	1.00000	1.00000	0.93809	0.96688	2.00000
(Chebyshev)	1.00000	1.00000	1.82192	0.49783	1.00000
1-dB ripple	0.95237	0.95237	1.00000	1.00000	1.95446

Finally, to obtain a set of reasonable values, we multiply the resistors by a factor of 10^5 and divide the capacitors by a factor of 10^5 (provided the factors are the same, the filter characteristics will not change). For good economy, we make $R_a = R_b = R_1 = R_2$. The component values of the filter are

$$R_1 = R_2 = R_a = R_b = 100 \text{ k}\Omega$$
$$C_1 = 927.375 \text{ pF}$$
$$C_2 = 1{,}213.949 \text{ pF}$$

Note: The designer might prefer to choose resistors of values of 10 kΩ to reduce bias current-offset effects, depending on the particular amplifier used. In this case the capacitors must be multiplied by a factor of 10, giving $C_1 = 0.00927$ μF and $C_2 = 0.01214$ μF.

In practice, the designer must use commercially available capacitors. A choice of a ceramic capacitor of 0.01 μF for C_1 and 0.01 μF or 0.015 μF for C_2 is suitable. Since these are not exactly equal to the calculated values, the cutoff frequency will be slightly different from the one specified. Rarely is this deviation of any consequence. If, for some reason, the precise cutoff frequency must be obtained, it is necessary to shunt the capacitors with trimming capacitors. In this case, the fixed capacitors must have smaller values than the ones computed (e.g., 0.008 μF for C_1 and 0.01 μF for C_2), and the trimmer capacitors will be adjusted to add the required amount of capacitance. If a precise cutoff frequency is required, trimmers (capacitor trimmers or resistance trimmers) must be added anyway, since commercial capacitors have typically $\pm20\%$ tolerance.

3.2.2 High-Pass Filters

The circuit diagram of a high-pass filter is given in Figure 3-2. Component values for the normalized filter are listed in Table 3-2. To see how to make use of this information, the reader must refer to the section on low-pass filters, and follow the same procedure. Select

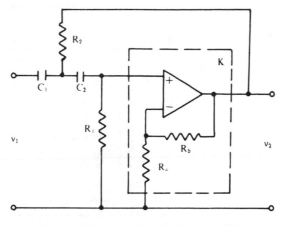

Figure 3-2   Second-order high-pass filter.

components from the table for the desired filter class, for $\omega_L = 1$; divide all capacitor values by the desired cutoff frequency; multiply all resistance values and divide all capacitance values by a suitable factor to obtain reasonable practical values of components and round the values off to commercially available values; use trimmers if necessary.

3.2.3 Band-Pass Filters

The circuit diagram of a band-pass filter is shown in Figure 3-3. A band-pass filter is characterized by the center frequency ω_C and by the Q of the filter. The Q of a filter is defined as the ratio of f_c (the center frequency) to the bandwidth (the difference $f_H - f_L$).

Table 3-3 lists values of K and of components for four different values of Q for the normalized (i.e., $\omega_c = 1$) filter. The following example demonstrates how to design a band-pass filter of the configuration of Figure 3-3 with the aid of Table 3-3.

TABLE 3-2   Components and Gain Values for the High-Pass Filter of Figure 3-2 for $\omega_1 = 1$ (R in ohms; C in farads)

Filter class	R_1	R_2	C_1	C_2	K
Maximally flat	1.00000	1.00000	1.41421	0.70711	2.00000
(Butterworth)	0.70711	1.41421	1.00000	1.00000	1.00000
3.01 dB at ω_H	1.00000	1.00000	1.00000	1.00000	1.58579
Equal ripple	1.00000	1.00000	1.42563	1.06356	2.00000
(Chebyshev)	0.71281	2.12707	1.00000	1.00000	1.00000
0.5-dB ripple	1.23134	1.23134	1.00000	1.00000	1.84222
Equal ripple	1.00000	1.00000	1.09772	1.00436	2.00000
(Chebyshev)	0.54586	2.00872	1.00000	1.00000	1.00000
1-dB ripple	1.05001	1.05001	1.00000	1.00000	1.00000

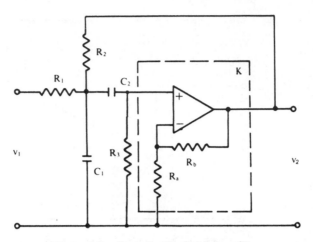

Figure 3-3   Second-order band-pass filter.

Example

Design a band-pass filter with a center frequency of 10 kHz and a Q of 5.

The circuit diagram is shown in Figure 3-3. From Table 3-3, we select for $Q = 5$ the values:

$$R_1 = 1.00000, \quad R_2 = 0.63439, \quad R_3 = 2.57630$$
$$C_1 = C_2 = 1.00000, \quad K = 2.00000$$

The value of K gives the gain of the amplifier:

$$K = 1 + \frac{R_a}{R_b}$$

We choose $R_a = R_b = 1.00000$, giving $K = 1 + 1 = 2$.

TABLE 3-3   Component Values for the Band-Pass Filter of Figure 3-3 for $\omega_c = 1$ rad/s (R in ohms; C in farads)

Q	R_1	R_2	R_3	C_1	C_2	K
2	1.41421	1.41421	1.41421	1.00000	1.00000	3.29284
	1.00000	0.74031	2.35078	1.00000	1.00000	2.00000
5	1.41421	1.41421	1.41421	1.00000	1.00000	3.71716
	1.00000	0.63439	2.57630	1.00000	1.00000	2.00000
10	1.41421	1.41421	1.41421	1.00000	1.00000	3.85858
	1.00000	0.60471	2.63587	1.00000	1.00000	2.00000
20	1.41421	1.41421	1.41421	1.00000	1.00000	3.92428
	1.00000	0.59076	2.69274	1.00000	1.00000	2.00000

The circuit of Figure 3-3 with component values listed above gives a filter with the desired $Q = 5$, but a center frequency $\omega_c = 1$. To obtain a center frequency $f_c = 10$ kHz (or $\omega_c = 2\pi10,000$), we divide the capacitor values by $2\pi10,000$:

$$C_1 = C_2 = \frac{1}{2\pi10,000} = 15.9 \ \mu F$$

Finally, to obtain reasonable values, we multiply all resistors by 10^4 and divide all capacitors by 10^4. The result is:

$$R_1 = R_a = R_b = 10 \ k\Omega; \ R_2 = 6.34 \ k\Omega; \ R_3 = 25.8 \ k\Omega$$
$$C_1 = C_2 = 1,590 \ pF$$

The nearest standard commercial values are:

$$R_1 = R_a = R_b = 10 \ k\Omega; \ R_2 = 6.2 \ k\Omega; \ R_3 = 27 \ k\Omega$$
$$C_1 = C_2 = 1,500 \ pF$$

These are slightly different from the computed values, causing f_c and Q to be slightly different from the specified values. If exact specifications must be met, trimmer capacitors must be added in shunt (parallel) with C_1 and C_2.

3.3 HIGH-GAIN AMPLIFIER FILTERS

The term *high-gain amplifier* refers to a conventional high-gain operational amplifier used directly as an element of the filter with the peripheral components, forming the required filter transfer characteristics. Unlike low-gain amplifier filters, this type often requires a spread in values of resistors. On the other hand, this type allows us to specify the gain of the filter through a choice of components, as given in the tables.

3.3.1 Low-Pass Filters

The circuit diagram of a second-order low-pass, high-gain amplifier filter is shown in Figure 3-4. Examples of component values for normalized maximally flat and equal-ripple filters are given in Table 3-4. To obtain a desired cutoff frequency of $\omega_H = 2\pi f_H$, we divide the values of all capacitors by ω_H. Then, to obtain a set of reasonable component values, we multiply the resistor values and divide the capacitance values by a single suitable factor just as we have demonstrated for the low-pass low-gain amplifier filter.

3.3.2 High-Pass Filters

The circuit diagram of a second-order high-pass, high-gain amplifier filter is shown in Figure 3-5. Examples of component values for normalized maximally flat and equal-ripple filters are given in Table 3-5. The use of this table to design filters is demonstrated

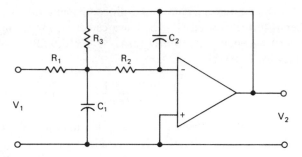

Figure 3-4   Second-order low-pass high-gain amplifier filter.

in the following example, which illustrates the required steps to arrive at a practical design.

Example

Design an equal-ripple (0.5-dB ripple) high-pass filter with a cutoff frequency of 5 kHz and a gain (at $f = \infty$) of 3.

The circuit configuration of the filter is given in Figure 3-5. The component values for the normalized filter are obtained from Table 3-5.

$$R_1 = 0.61099 \qquad R_2 = 7.44475$$
$$C_1 = 1.00000 \qquad C_2 = 1.00000 \qquad C_3 = 0.33333$$

To obtain the cutoff frequency of $f_1 = 5$ kHz, we divide all the capacitors by $\omega_L = 2\pi f_L = 2\pi 5,000$. This gives

TABLE 3-4   Component Values for the Low-Pass Filter of Figure 3-4 for $\omega_H = 1$ (R in ohms; C in farads)

Filter class	DC gain	R_1	R_2	R_3	C_1	C_2
Maximally flat	1	1.00000	1.00000	1.00000	2.12133	0.47140
(Butterworth)	3	1.00000	1.00000	3.00000	1.64991	0.20203
3.01 dB at ω_H	5	1.00000	1.00000	5.00000	1.55563	0.12857
Equal ripple	1	1.00000	1.00000	0.65954	2.46644	0.40544
(Chebyshev)	3	1.00000	1.00000	1.97863	1.75741	0.18967
0.5-dB ripple	5	1.00000	1.00000	3.29771	1.61560	0.12379
Equal ripple	1	1.00000	1.00000	0.90702	2.82628	0.35382
(Chebyshev)	3	1.00000	1.00000	2.72106	2.15672	0.15456
1-dB ripple	5	1.00000	1.00000	4.53508	2.02280	0.09887

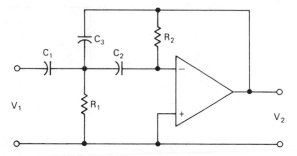

Figure 3-5   Second-order high-pass high-gain amplifier filter.

$$C_1 = C_2 = 1.00000/2\pi5,000 = 31.83099\ \mu F$$
$$C_3 = 0.33333/2\pi5,000 = 10.61022\ \mu F$$

Finally, to obtain a set of reasonable values, we multiply the resistor values by a factor of 10^4 and divide the capacitor values by the same factor. The component values of the filter then are

$$R_1 = 6.1099\ k\Omega \qquad R_2 = 74.4475\ k\Omega$$
$$C_1 = C_2 = 3183.099\ \mu F$$
$$C_3 = 1061.022\ pF$$

In practice we use the closest commercially available components and add trimmer pots if precise values are needed. Usually, such a degree of precision for filters is not required.

TABLE 3-5   Component Values for the High-Pass Filter of Figure 3-5 for $\omega_L = 1$ (R in ohms; C in farads)

Filter class	Gain at $f = \infty$	R_1	R_2	C_1	C_2	C_3
Maximum float	1	0.47140	2.12132	1.00000	1.00000	1.00000
(Butterworth)	3	0.60609	4.94976	1.00000	1.00000	0.33333
3.01 dB at ω_L	5	0.64282	7.77817	1.00000	1.00000	0.20000
Equal ripple	1	0.47521	3.19061	1.00000	1.00000	1.00000
(Chebyshev)	3	0.61099	7.44475	1.00000	1.00000	0.33333
0.5-dB ripple	5	0.64801	11.69889	1.00000	1.00000	0.20000
Equal ripple	1	0.36591	3.01308	1.00000	1.00000	1.00000
(Chebyshev)	3	0.47045	7.03051	1.00000	1.00000	0.33333
1-dB ripple	5	0.49897	11.04795	1.00000	1.00000	0.20000

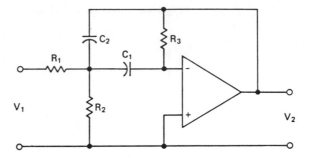

Figure 3-6   Second-order band-pass, high-gain amplifier filter.

3.3.3 Band-Pass Filters

The circuit diagram of a second-order band-pass, high-gain amplifier filter is shown in Figure 3-6. Several sets of component values for different Q-factors and resonant frequency gains are given in Table 3-6. The Q-factor defines the sharpness of the filter. It specifies the ratio of the resonant frequency to the bandwidth (i.e., the frequency band between the low and high 3-dB points). The higher the Q-factor, the wider the spread of the resistor values, making high Q-factors for this design configuration impractical. The component values in Table 3-6 are for normalized filters. In filter design, these values are changed to obtain the desired resonance frequency and reasonable resistor and capacitor values by going through steps similar to those taken in the example for high-pass filters in the previous paragraph, or in the example for band-pass filters using low-gain amplifiers given earlier in this chapter.

TABLE 3-6   Component Values for the Band-Pass Filter of Figure 3-6 for $\omega_o = 1$

Q	Gain at resonance	R_1	R_2	R_3	C_1	C_2
	1	2.00000	0.28571	4.00000	1.00000	1.00000
2	2	1.00000	0.33333	4.00000	1.00000	1.00000
	4	0.50000	0.50000	4.00000	1.00000	1.00000
	6	0.33333	1.00000	4.00000	1.00000	1.00000
	1	5.00000	0.10204	10.00000	1.00000	1.00000
5	2	2.50000	0.10417	10.00000	1.00000	1.00000
	4	1.25000	0.10870	10.00000	1.00000	1.00000
	6	0.83333	0.11364	10.00000	1.00000	1.00000
	1	8.00000	0.12698	16.00000	1.00000	1.00000
8	2	4.00000	0.12903	16.00000	1.00000	1.00000
	4	2.00000	0.13333	16.00000	1.00000	1.00000
	6	1.33333	0.12766	16.00000	1.00000	1.00000

3.4 CHOICE BETWEEN HIGH-GAIN AND LOW-GAIN AMPLIFIER FILTERS

Low-gain and high-gain amplifier filters can be designed to give similar frequency-response characteristics. The choice is dictated by other considerations. High-gain amplifier filters require fewer components, and the desired gain is obtained by the choice of component values. On the other hand, low-gain amplifier filters allow for a large number of identical component values for a given filter, resulting in less room for error and lower production costs in large quantity runs.

4

Designing Third-Order Filters

4.1 INTRODUCTION

Third-order filters are perhaps the most widely used filters. Like first- and second-order filters, they require only a single operational amplifier, yet in the stop band they provide a steep slope of -60 dB/dec. The tables in this chapter give filter component values. Equations given make it possible to compute the exact attenuation in the stop band.

4.2 LOW-PASS FILTERS

The circuit diagram of a low-pass third-order filter is shown in Figure 4-1. Component values to achieve particular characteristics are listed in Table 4-1. The maximally flat

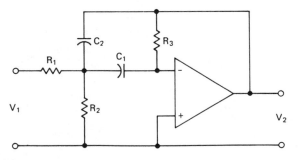

Figure 4-1 Third-order low-pass filter. (Reprinted from *Electronics,* November 11, 1976. Copyright © McGraw-Hill, Inc., 1976.)

TABLE 4-1   Component Values for the Filter of Figure 4-1 for $\omega_H = 1$ rad/s (R in ohms; C in farads)

Filter class	R_1	R_2	R_3	C_1	C_2	C_3
Maximally flat (Butterworth) 3.01 dB at ω_2	1.00000	1.00000	1.00000	0.20245	3.5465	1.3926
Equal ripple (Chebyshev) Ripple (dB)						
0.001	1.00000	1.00000	1.00000	0.07130	2.5031	0.8404
0.03	1.00000	1.00000	1.00000	0.07736	3.3128	1.0325
0.10	1.00000	1.00000	1.00000	0.09691	4.7921	1.3145
0.30	1.00000	1.00000	1.00000	0.08582	7.4077	1.6827
1.00	1.00000	1.00000	1.00000	0.05872	14.784	2.3444

Reprinted from *Electronics*, November 11, 1976; Copyright © McGraw-Hill, Inc., 1976

filter frequency response curve has an asymptotic slope in the stop band of -60 dB/dec (-18 dB/oct). The exact attenuation at any frequency ω is given by

$$\frac{V_2}{V_1} = \frac{1}{(1 + \omega^6)^{1/2}} \tag{4-1}$$

The exact attenuation for the equal-ripple filters in the stop band is given by

$$\frac{V_2}{V_1} = \frac{1}{[1 + \epsilon^2(4\omega^3 - 3\omega)^2]^{1/2}} \tag{4-2}$$

where ϵ^2 is the *ripple factor*. Values of the ripple factor for the ripple values in Table 4-1 are given in Table 4-2.

TABLE 4-2   Ripple Factors for Equal-Ripple Filters

Ripple (dB)	Ripple factor (ϵ^2)
0.01	0.00230524
0.03	0.00693167
0.1	0.023293
0.3	0.0715193
1.0	0.255925

Reprinted from *Electronics*, November 11, 1976; Copyright © McGraw-Hill, Inc., 1976

Example

Design a low-pass equal-ripple filter with a ripple of 0.1 dB and a cutoff frequency of 5 kHz. The filter configuration is shown in Figure 4-1. The component values for the normalized filter are obtained from Table 4-1.

$$R_1 = R_2 = R_3 = 1 \ \Omega; \ C_1 = 0.09691 \ \text{F}; \ C_2 = 4.7921 \ \text{F};$$
$$C_3 = 1.3145 \ \text{F}$$

To have a cutoff frequency of $\omega_H = 2\pi 5 \times 10^3 = 31.416 \times 10^3$ rad/s, we divide all the capacitors by 31.416×10^3 giving

$$C_1 = 3.085 \ \mu\text{F}; \ C_2 = 152.537 \ \mu\text{F}; \ C_3 = 41.842 \ \mu\text{F}$$

Finally, to obtain practical component values, we multiply all the resistor values by 10^4 and divide all the capacitor values by the same factor, giving

$$R_1 = R_2 = R_3 = 10 \ \text{k}\Omega$$
$$C_1 = 308.5 \ \text{pF}; \ C_2 = 0.0153 \ \mu\text{F}; \ C_3 = 0.0048 \ \mu\text{F}$$

The frequency response of the filter can be obtained with the aid of Equation 4-2. This equation is for the normalized filter. To use it for the denormalized filter, ω must be replaced with f/f_H, where f_H is the passband cutoff frequency.

$$\frac{V_2}{V_1} = \frac{1}{\left[1 + 0.023293 \left(4 \ \dfrac{f^3}{1.25 \times 10^{11}} -3 \ \dfrac{f}{5 \times 10^3} \right)^2 \right]^{1/2}}$$

where we replaced ϵ^2 with 0.023293 as given in Table 4-2 for 0.1-dB ripple. For an illustration, we compute the attenuation at 5,000 Hz and at 7,000 Hz.
    At a frequency of 5,000 Hz:

$$\frac{V_2}{V_1} = \frac{1}{[1 + 0.023293 \ (4 - 3)^2]^{1/2}} = \frac{1}{(1.023293)^{1/2}} = 0.98855$$

In decibels,

$$20 \ \log_{10} 0.9886 = -0.1000 \ \text{dB}$$

That is, at 5 kHz the gain is down by 0.1 dB from the DC gain. At a frequency of 7,000 Hz:

$$\frac{V_2}{V_1} = \frac{1}{\left[1 + 0.023293 \left(4 \times \dfrac{3.43 \times 10^{11}}{1.25 \times 10^{11}} -3 \times \dfrac{7 \times 10^3}{5 \times 10^3} \right)^2 \right]^{1/2}} = 0.6951$$

In decibels,

$$20 \ \log_{10} 0.6951 = -3.159 \ \text{dB}$$

Thus, at 7 kHz the gain is down by 3.159 dB from the DC gain. Additional points can be computed to plot the complete frequency-response curve of the filter.

4.3 HIGH-PASS FILTERS

The circuit configuration for a high-pass filter is shown in Figure 4-2. The component values for the normalized filter are given in Table 4-3. If the cutoff frequency must be ω_L

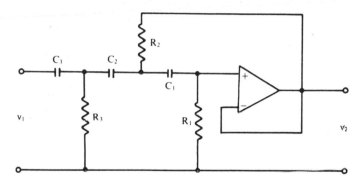

Figure 4-2 Third-order high-pass filter. (Reprinted from *Electronics,* November 11, 1976. Copyright © McGraw-Hill, Inc., 1976.)

$= 2\pi f_L$ rather than $\omega = 1$, all the capacitors must be divided by ω_L, just as was done in the example for the low-pass filter. Finally, to obtain practical component values, all the resistors are multiplied by a given factor and all the capacitors are divided by the same factor. This was also demonstrated for the low-pass filter.

TABLE 4-3   Component Values for the Filter of Figure 4-2 for $\omega_L = 1$ rad/s (R in ohms; C in farads)

Filter class	R_1	R_2	R_3	C_1	C_2	C_3
Maximally flat (Butterworth) 3.01 dB at ω_1	4.93949	0.28194	0.71808	1.00000	1.00000	1.00000
Equal ripple (Chebyshev) Ripple (dB)						
0.01	10.9130	0.39450	1.18991	1.00000	1.00000	1.00000
0.03	10.09736	0.30186	0.96852	1.00000	1.00000	1.00000
0.10	10.3188	0.20868	0.76075	1.00000	1.00000	1.00000
0.30	11.65230	0.13499	0.59428	1.00000	1.00000	1.00000
1.00	17.0299	0.06764	0.42655	1.00000	1.00000	1.00000

Reprinted from *Electronics*, November 11, 1976; Copyright © McGraw-Hill, Inc., 1976

The frequency responses for the normalized low-pass maximally flat and equal-ripple filters are given respectively by Equations 4-1 and 4-2. For the normalized high-pass filters, ω in these equations must be replaced with $1/\omega$. For the denormalization, high-pass filters ω in Equations 4-1 and 4-2 must be replaced with f_L/f, where f_L is the cutoff frequency.

4.4 BAND-PASS FILTERS

Band-pass filters are realized by cascading low-pass and high-pass filters. To prevent interaction between the filters, one connects a voltage follower between the two stages.

5

Designing Fourth-Order Filters

5.1 INTRODUCTION

Fourth-order filters consist of cascaded second-order filters. Since there is interaction between the two stages, the total filter must be designed as a single entity rather than as a cascade connection of two separate entities. The component values for the two sections are not the same, and the order in which the sections are connected is important. The tables in this chapter provide all the required information; the worked-out numerical examples illustrate the use of the tables.

5.2 LOW-PASS FILTERS

The circuit diagram of a fourth-order low-pass filter is shown in Figure 5-1. The values of the components are determined with the aid of Table 5-1. The table lists component values for a maximally flat response and for an equal 3-dB ripple response. Many different combinations of components yield the right characteristics. In Table 5-1 we list two possible combinations of components for each stage of each class. Either set of components of the first stage can be combined with either set of the second stage of the same class. Components for the first set of each stage are selected so that either both resistors or both capacitors and as many other components as possible are equal to unity. Components for the second set are chosen so that the gain of the amplifier is equal to an integer. This makes it possible to use two resistors of the same value for the amplifier. In particular, a gain of 2 results in resistance values (of R_{a1} and R_{b1} for the first stage and R_{a2} and R_{b2} for the second stage) of unity, since the gain is given by unity plus the ratio of the resistances.

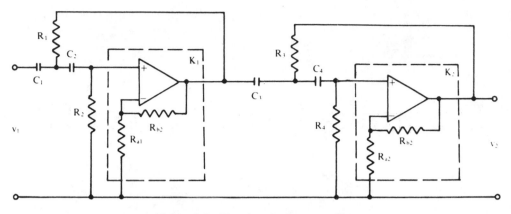

Figure 5-1   Fourth-order low-pass filter.

In one case (first stage of the low-pass equal-ripple filter), the gain is set equal to 5. A lower integer in this case results in negative values of other components, and is therefore physically unrealizable. We shall illustrate the use of the tables in the following example.

TABLE 5-1   Component Values for the Filter of Figure 5-1 for $\omega_H = 1$ (R in ohms; C in farads)

Filter class	Stage	1	R_1	R_2	C_1	C_2	K_1
		2	R_3	R_4	C_3	C_4	K_2
Maximally flat		1	1.0000	1.0000	1.0000	1.0000	1.1522
			1.0000	0.5412	1.0000	1.8478	2.0000
	Stage						
(Butterworth)		2	1.0000	1.0000	1.0000	1.0000	2.2346
			1.0000	1.3065	1.0000	0.7654	2.0000
Equal ripple (Chebyshev)		1	1.0000	5.1020	1.0000	1.0000	5.0041
			1.0000	5.1225	1.0000	0.4960	5.0000
	Stage						
Ripple: 3 dB		2	1.0000	1.1073	1.0000	1.0000	2.1986
			1.0000	13.6032	1.0000	0.0816	2.0000

Example

Design a fourth-order maximally flat filter with a cutoff frequency of 800 Hz. The circuit diagram is shown in Figure 5-1. From Table 5-1 we choose for the first stage $R_1 = R_2 = 1\ \Omega$, $C_1 = C_2 = 1$ F. The gain of the amplifier is 1.1522. We let $R_{a1} = 1\ \Omega$, and compute R_{b1}:

$$K = 1 + \frac{R_{b1}}{R_{a1}} = 1.1522 \qquad \frac{R_{b1}}{R_{a1}} = 1.1522 - 1 = 0.1522$$

$$R_{b1} = 0.1522\ R_{a1} = 0.1522\ \Omega$$

A filter with these components has an upper cutoff frequency $\omega_H = 1$ rad/s.

To shift the cutoff frequency to the specified 800-Hz value, we divide all the capacitor values by $2\pi800$. This yields

$$C_1 = C_2 = \frac{1}{2\pi800} = 0.000199 \text{ F}$$

Finally, to obtain reasonable resistor and capacitor values, we multiply all resistor values by 10^4 and divide all capacitor values by 10^4:

$$R_1 = R_2 = R_{a1} = 10 \text{ k}\Omega; R_{b1} = 1.522 \text{ k}\Omega$$
$$C_1 = C_2 = 0.0199 \text{ }\mu\text{F}$$

Commercially available values are:

$$R_1 = R_2 = R_{a1} = 10 \text{ k}\Omega; R_{b1} = 1.5 \text{ k}\Omega$$
$$C_1 = C_2 = 0.022 \text{ }\mu\text{F}$$

The design of the second stage is similar to that of the first stage. From Table 5-1 select

$$R_3 = R_4 = 1 \text{ }\Omega; C_3 = C_4 = 1 \text{ F}; K = 2.2346$$

choosing $R_{a2} = 1 \text{ }\Omega$ gives $R_{b2} = 1.2346 \text{ }\Omega$.

Divide the capacitor values by $2\pi800$:

$$C_3 = C_4 = 0.000199 \text{ F}$$

Multiply all resistors by 10^4 and divide all capacitors by 10^4:

$$R_3 = R_4 = R_{a2} = 10 \text{ k}\Omega; R_{b2} = 12.346 \text{ k}\Omega$$
$$C_3 = C_4 = 0.0199 \text{ }\mu\text{F}$$

Commercially available values are:

$$R_3 = R_4 = R_{a2} = 10 \text{ k}\Omega; R_{b2} = 12 \text{ k}\Omega$$
$$C_3 = C_4 = 0.022 \text{ }\mu\text{F}$$

The commercially available components are not all equal in value to the computed value. Furthermore, components have tolerances (e.g., resistors ±5%; capacitors ±20%). This results in a cutoff frequency different from the specified value. If the specification is critical, the filter must be trimmed using resistor and/or capacitor trimmers.

5.3 HIGH-PASS FILTERS

The circuit diagram of a fourth-order high-pass filter is shown in Figure 5-2. The values of the components are determined with the aid of Table 5-2. The table lists component values for a maximally flat response and for an equal 3-dB ripple response. Many different combinations of components yield the right characteristics. In Table 5-2 two possible

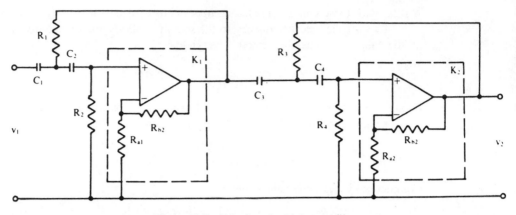

Figure 5-2   Fourth-order high-pass filter.

combinations of components are listed for each stage of each class. Either set of components of the first stage can be combined with either set of the second stage of the same
class. Components for the first set of each stage are selected so that either both resistors or
both capacitors and as many other components as possible are equal to unity. Components
for the second set are chosen so that the gain of the amplifier is equal to an integer. This
makes it possible to use two resistors of the same value for the amplifier. In particular, a
gain of 2 results in resistance values (of R_{a1} and R_{b1} for the first stage and R_{a2} and R_{b2} for
the second stage) of unity, since the gain is given by unity plus the ratio of the resistances.

TABLE 5-2   Component Values for the Filter of Figure 5-2 for $\omega_L = 1$ (R in ohms; C in farads)

Filter class	Stage	1	R_1	R_2	C_1	C_2	K_1
		2	R_3	R_4	C_3	C_4	K_2
Maximally flat		1	1.0000	1.0000	1.0000	1.0000	1.1522
			1.0000	0.5412	1.0000	1.8478	2.0000
	Stage						
(Butterworth)		2	1.0000	1.0000	1.0000	1.0000	2.2346
			1.0000	1.3065	1.0000	0.7654	2.0000
Equal ripple		1	1.0000	1.0000	1.0000	0.1960	0.9808
(Chebyshev)			1.0000	0.1613	1.0000	1.2152	2.0000
	Stage						
Ripple: 3 dB		2	1.0000	1.0000	1.0000	0.9031	2.6358
			1.0000	0.2960	1.0000	0.2673	2.0000

Example

Design a fourth-order equal-ripple (3-dB) filter with a cutoff frequency
of 2 kHz. The circuit diagram is shown in Figure 5-2. From Table 5-2 we
choose for the first stage

$$R_1 = 1.0000 \ \Omega, R_2 = 0.1613 \ \Omega$$
$$C_1 = 1.0000 \ F, C_2 = 1.2152 \ F$$
$$K_1 = 2.0000$$

For the second stage we choose

$$R_3 = 1.0000 \ \Omega, R_4 = 0.2960 \ \Omega$$
$$C_3 = 1.0000 \ F, C_4 = 0.2673 \ F$$
$$K_2 = 2.0000$$

Resistors $R_{a1}, R_{b1}, R_{a2}, R_{b2}$ are computed from $1 + R_{b1}/R_{a1} = K_1 = 2$ and $1 + R_{b2}/R_{a2} = K_2 = 2$. A good choice is $R_{a1} = R_{b1} = R_{a2} = R_{b2} = 1 \ \Omega$. This gives a large number of resistors with an identical value. Divide the capacitors by $2\pi \times 2,000$. This gives

$$C_1 = C_3 = 0.0000796 \ F$$
$$C_2 = 0.0000967 \ F$$
$$C_4 = 0.0000213 \ F$$

Multiply resistor values by 10^5 and divide capacitor values by 10^5.

$$R_1 = R_3 = R_{a1} = R_{b1} = R_{a2} = R_{b2} = 100 \ k\Omega$$
$$R_2 = 16.13 \ k\Omega \ R_4 = 29.60 \ k\Omega$$
$$C_1 = C_3 = 796 \ pF, C_2 = 967 \ pF, C_4 = 213 \ pF$$

The designer now chooses appropriate commercially available components and trims the circuit if necessary.

5.4 BAND-PASS FILTERS

Band-pass filters are realized by cascading low-pass and high-pass filters. Interaction between the filters is prevented by interfacing them with a voltage follower.

6

Designing Higher-Order Filters

6.1 INTRODUCTION

Higher-order filters provide very steep rolloff in the stop band. It would be too cumbersome and lengthy to provide tables for all the components of the many possible filters. Instead, we shall provide one table with expressions from which all the component values can be computed. Higher-order filters are designed by cascading lower-order filter configurations with component values obtained from the table in this section. The lower-order filters forming the links of the higher-order filters will not be identical. An even-order filter consists of cascaded second-order filters such as shown in Figure 3-1 for low-pass filters or in Figure 3-2 for high-pass filters. An odd-order filter consists of second-order filters as above and a first-order filter as shown in Figure 2-1 for low-pass filters or in Figure 2-2 for high-pass filters. The first-order filter can be replaced with a simple passive RC network as shown in Figures 6-1 and 6-2. The maximum gain of these filters in the passband is unity, whereas the gain of the active filters of Figures 2-1 and 2-2 is given by the ratio of the

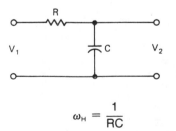

$$\omega_H = \frac{1}{RC}$$

Figure 6-1   Passive first-order low-pass filter.

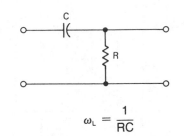

$$\omega_L = \frac{1}{RC}$$

Figure 6-2   Passive first-order high-pass filter.

feedback resistor to the input resistor (R_2/R_1). When the passive filters of Figures 6-1 and 6-2 are used, they must constitute the last stage of the filter to avoid loading problems. This assumes that the filter looks into a high-impedance circuit or device. If this is not the case, the filter must be followed with a voltage follower for buffering. If a buffer is used, then of course the stage can be used anywhere in the chain of the cascaded filter, and it would be just as appropriate to use an active first-order filter in the first place. A property of active filters is their low-output impedance because of the low-output impedance of opamps, which makes loading by the next stage negligible.

6.2 LOW-PASS FILTERS

The expressions for the second-order filter links in Table 6-1 have the form

$$x^2 + a_1 x + a_0$$

The constants a_1 and a_0 are used to design the several stages of the normalized filter. This is followed by two steps that we have introduced earlier, namely, dividing the capacitor values by the desired cutoff frequency ω_H and then multiplying the resistor values, and dividing the capacitor values by a chosen constant to obtain a set of reasonable component values.

For the normalized filter (see symbols in Figure 3-1)

$$\frac{1}{R_1 C_1} + \frac{1}{R_2 C_1} + \frac{1 - K}{R_2 C_2} = a_1 \tag{6-1}$$

and

$$\frac{1}{R_1 R_2 C_1 C_2} = a_0 \tag{6-2}$$

We have two equations with five unknowns: R_1, R_2, C_1, C_2, and K. Therefore, we can choose arbitrarily any three values and solve for the remaining two. For example, we can choose $R_1 = R_2 = C_1 = 1$ (ohms and farads). From the second equation above, we then have $C_2 = 1/a_0$. We substitute this value for C_2 in the first equation above and solve for the remaining unknown K. Or, we can choose $K = 2$, $R_1 = R_2 = 1$. This results in $R_a = R_b$ (Figure 3-1), which will give good temperature tracking. Using the two equations

TABLE 6-1 Expressions for Higher-Order Filters (Normalized filters)

Filter class	Filter order	1st Stage	2nd Stage	3rd Stage	4th Stage	5th Stage
Equal ripple (Chebyshev) 3-dB ripple	5	$x^2 + 1.6180x + 1$	$x^2 + 1.6180x + 1$	$x + 1$		
	6	$x^2 + 1.9318x + 1$	$x^2 + 1.4142x + 1$	$x^2 + 0.5176x + 1$		
	7	$x^2 + 1.8020x + 1$	$x^2 + 1.2470x + 1$	$x^2 + 0.4450x + 1$	$x + 1$	
	8	$x^2 + 1.9616x + 1$	$x^2 + 1.6630x + 1$	$x^2 + 1.1112x + 1$	$x^2 + 0.3902x + 1$	
	9	$x^2 + 1.8794x + 1$	$x^2 + 1.5320x + 1$	$x^2 + 1.0000x + 1$	$x^2 + 0.3474x + 1$	$x + 1$
	10	$x^2 + 1.9740x + 1$	$x^2 + 1.7820x + 1$	$x^2 + 1.4142x + 1$	$x^2 + 0.9080x + 1$	$x^2 + 0.3128x + 1$
Maximally flat (Butterworth)	5	$x^2 + 0.2872x + 0.3770$	$x^2 + 0.1096x + 0.9360$	$x + 0.1775$		
	6	$x^2 + 0.2854x + 0.0888$	$x^2 + 0.2090x + 0.5219$	$x^2 + 0.0764x + 0.9548$		
	7	$x^2 + 0.2288x + 0.2042$	$x^2 + 0.1575x + 0.6273$	$x^2 + 0.0564x + 0.9685$	$x + 0.1265$	
	8	$x^2 + 0.2170x + 0.6503$	$x^2 + 0.1340x + 0.3269$	$x^2 + 0.1230x + 0.7035$	$x^2 + 0.0432x + 0.9742$	
	9	$x^2 + 0.1848x + 0.1267$	$x^2 + 0.1506x + 0.4224$	$x^2 + 0.0982x + 0.7547$	$x^2 + 0.2342x + 0.4796$	$x + 0.983$
	10	$x^2 + 0.1746x + 0.0323$	$x^2 + 0.1576x + 0.2140$	$x^2 + 0.1250x + 0.5079$	$x^2 + 0.8825x + 0.8018$	$x^2 + 0.0276x + 0.9833$

above, we solve for C_1 and C_2. Clearly, many combinations of component values are possible.

For the first-order filter link in Table 6-1 we have the form

$$x + a_0$$

For the circuit configuration of Figure 6-1 we have

$$\frac{1}{RC} = a_0 \tag{6-3}$$

We then choose either R or C and compute the value of the remaining component from this equation. If we use the filter configuration of Figure 2-1, then R and C represent R_2 and C_2 in Figure 2-1, respectively. The value of R_1 in Figure 2-1 is chosen arbitrarily. It does not affect the frequency response of the filter and provides a DC gain R_2/R_1 for the stage.

Example

Design a sixth-order maximally flat low-pass filter with a high-frequency cutoff $f_H - 4$ kHz.

We note in Table 6-1 that for the maximally flat filter the coefficients a_0 of all the stages are equal to unity. Choosing for all stages $R_1 = R_2 = C_1 = 1$, we have from Equation 6-2, $C_2 = 1$, and from Equation 6-1,

$$3 - K = a_1$$
$$K = 3 - a_1$$

or, for the first stage,

$$a_1 = 1.9318$$

giving

$$K = 3 - 1.9318 = 1.0682$$

or

$$1 + R_b/R_a = 1.0682$$

or

$$R_b/R_a = 1.0682 - 1 = 0.0682$$

Letting $R_a = 1$, we have $R_b = 0.0682$.

By a similar procedure, for the second stage,

$$K = 3 - 1.4142 = 1.5858$$
$$R_a = 1 \qquad R_b = 0.5858$$

and for the third stage,

$$K = 3 - 0.5176 = 2.4824$$
$$R_a = 1 \qquad R_b = 1.4824$$

Summary for the normalized filter:

1. All Rs $= 1\ \Omega$, except $R_b = 0.0682$, 0.5858, and 1.4824, for the first, second, and third stages, respectively.

2. All Cs $= 1$ F.

To get the desired cutoff frequency $f_H = 4$ kHz, we divide all Cs by $2\pi 4{,}000$, giving

$$C = 1/2\pi 4000 = 39.7887\ \mu\text{F}$$

Finally, to get reasonable values for the resistors and capacitors, we multiply all the resistors by 10^4 and divide all the capacitors by 10^4. This gives

all R_1s, R_2s, and R_as $= 10$ kΩ
$R_b = 682$, $5{,}858$, $14{,}824\ \Omega$ for the first,
second, and third stages, respectively.
all Cs $= 3979$ pF

The complete filter is shown in Figure 6-3. Again, as in the previous cases, commercially available components must be used and capacitor or resistor trimmers added if necessary.

6.3 HIGH-PASS FILTERS

The design of high-pass filters also makes use of Table 6-1. But before the a values for the several stages can be obtained, certain computations with these expressions must be made. Specifically, the x in the expressions is replaced with $1/y$. Thus,

$$x^2 + a_1 x + a_0 \qquad x + a_0$$

become

$$\frac{1}{y^2} + \frac{a_1}{y} + a_0 \qquad \frac{1}{y} + a_0$$

We multiply the first expression by y^2 and the second by y:

$$a_0 y^2 + a_1 y + 1 \qquad a_0 y + 1$$

Next, we divide the expressions by a_0:

$$y^2 + \frac{a_1}{a_0} y + \frac{1}{a_0} \qquad y + \frac{1}{a_0}$$

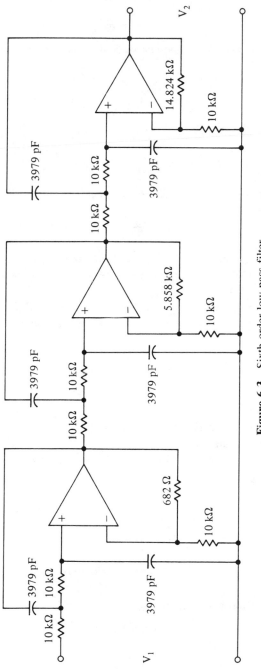

Figure 6-3 Sixth-order low-pass filter.

Finally, letting $\dfrac{a_1}{a_0} = b_1$ and $\dfrac{1}{a_0} = b_0$, we have

$$y^2 + b_1 y + b_0 \qquad y + b_0$$

For a second-order link, the relation between the component values and b_1 and b_0 is given by (see Figure 5-2)

$$\frac{1}{R_1 C_1} + \frac{1}{R_1 C_2} + \frac{1 - K}{R_2 C_1} = b_1 \tag{6-4}$$

and

$$\frac{1}{R_1 R_2 C_1 C_2} = b_0 \tag{6-5}$$

For the first-order link (Figure 6-2), the relation is

$$\frac{1}{RC} = b_0 \tag{6-6}$$

If the active filter of Figure 2-2 is used instead of the passive filter of Figure 6-2, the R and C are replaced with R_1 and C_1, respectively. Resistor R_2 can assume any value. The gain at $f = \infty$ of this link is given by the ratio R_2/R_1.

Note that for the maximally flat (Butterworth) frequency response, in Table 6-1, $a_0 = 1$. Consequently, $b_1 = a_1$ and $b_0 = a_0$. For equal-ripple (Chebyshev) response, the relationship is

$$b_1 = \frac{a_1}{a_0} \text{ and } b_0 = \frac{1}{a_0}$$

Example

Design a fifth-order equal-ripple (3-dB ripple) high-pass filter with a low-frequency cutoff at $f_L = 1$ kHz.

From Table 6-1 we have for the first stage $a_1 = 0.2872$ and $a_0 = 0.3770$. Hence,

$$b_1 = \frac{a_1}{a_0} = 0.7618 \qquad b_0 = \frac{1}{a_0} = 2.6525$$

Let $R_1 = C_1 = C_2 = 1$. From Equation 6-5,

$$R_2 = \frac{1}{b_0} = 0.3770$$

Substituting into Equation 6-4 and solving for K,

$$K = (1 + 1 - 0.7618)0.3770 + 1 = 1.4668$$

But

$$1 + \frac{R_b}{R_a} = K$$

Let $R_a = 1$, then $R_b = 0.4668$.

Summary for first stage

$$R_1 = R_a = 1 \ \Omega; R_2 = 0.3770 \ \Omega; R_b = 0.4668 \ \Omega; C_1 = C_2 = 1 \ \text{F}$$

For the second stage, from Table 6-1, $a_1 = 0.1096$ and $a_0 = 0.9360$ giving $b_1 = 0.1171$ and $b_0 = 1.0684$.

Following the same procedure used for the first stage, we have:

Summary for second stage

$$R_1 = R_a = 1 \ \Omega; R_2 = 0.9360 \ \Omega; R_b = 1.7624 \ \Omega; C_1 = C_2 = 1 \ \text{F}$$

The third stage is a first-order filter. From Table 6-1, $a_0 = 0.1775$, from which

$$b_0 = \frac{1}{a_0} = 5.6338$$

From equation 6-6, $RC = 5.6338$. Let $C = 1$ F. Then $R = 5.6338 \ \Omega$.

To provide the cutoff frequency of $f_L = 1000$ Hz, all the capacitor values are divided by $2\pi 1000$ giving for all Cs the value 159.1549 µF.

Finally, to have reasonable component values, we multiply all the resistors by 10^5 and divide all the capacitors by 10^5. The component values are then as follows:

First stage:

$$R_1 = R_a = 100 \ \text{k}\Omega; R_2 = 37.70 \ \text{k}\Omega; R_b = 46.68 \ \text{k}\Omega$$
$$C_1 = C_2 = 1{,}591.55 \ \text{pF}.$$

Second stage:

$$R_1 = R_a = 100 \ \text{k}\Omega; R_2 = 93.60 \ \text{k}\Omega; R_b = 176.24 \ \text{k}\Omega$$
$$C_1 = C_2 = 1{,}591.55 \ \text{pF}.$$

Third stage:

$$R = 563.38 \ \text{k}\Omega; \qquad C = 1{,}591.55 \ \text{pF}$$

Examining these values, we find that the third stage will have better values if we divide the resistor values by 100 and multiply the capacitor values by the same factor. This gives

Third stage:

$$R = 5.63 \ \text{k}\Omega; \qquad C = 0.1592 \ \text{µF}$$

The circuit diagram of the filter is shown in Figure 6-4. As in all previous cases, in practice, commercially available resistor and capacitor values must be used and trimmers added if needed.

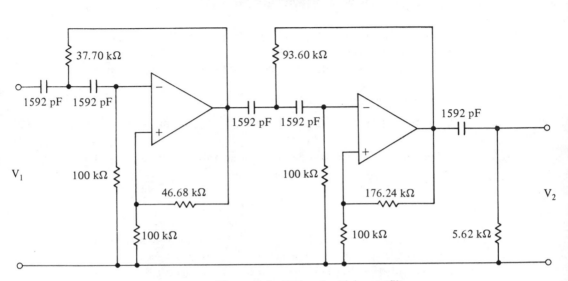

Figure 6-4   Fifth-order high-pass filter.

6.4 BAND-PASS FILTERS

Low-Q high-order band-pass filters can be obtained by cascading low-pass filters with high-pass filters. Usually band-pass filters are designed to obtain high selectivity, which requires high-Q filters. Such filters are discussed in the next section.

6.5 HIGH-Q BAND-PASS FILTERS

The Q of a band-pass filter is defined as the ratio of the center frequency to the bandwidth (i.e., the frequency range between the 3-dB points). A high-Q represents therefore a "sharp" or highly selective filter. The band-pass filters discussed in previous sections are low-Q, or broadband, filters (Q of the order of 10). The circuit diagram shown in Figure 6-5 is a high-Q filter (Qs up to several hundreds). It is known as a *state-variable filter*. A design procedure can be formulated by letting, for the normalized filter,

$$R_1 = R_2 = R_3 = R_5 = R_6 = 1 \ \Omega$$
and
$$C_1 = C_2 = 1 \ \text{F}$$

The remaining resistor, R_4, is then computed from

$$R_4 = \frac{2}{\dfrac{1}{Q} - \dfrac{1}{A_1} - \dfrac{1}{A_2}}$$

where A_1 and A_2 are the *open-loop* gains of the amplifiers as shown in Figure 6-5. For gains in the hundreds of thousands, and Qs in the hundreds, we have

$$R_4 \simeq 2Q$$

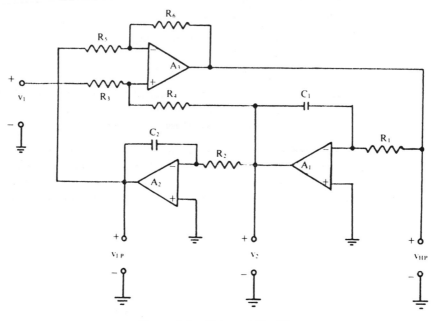

Figure 6-5   State variable filter.

The open-loop gain A_3 must be high, but does not enter the calculation. The gain at resonance ($\omega_0 = 1$ rad/s for the normalized filter) is R_4/R_1. Since R_1 is 1 Ω, the gain is numerically equal to R_4.

We shall illustrate the use of these equations and show how to design for a desired resonance filter with practical component values in a design example.

Example

Design a high-Q band-pass filter with a center, or resonance, frequency at 5 kHz and a Q of 250.

We select amplifiers with gains of 10,000, resistors (except R_4) of 1 Ω, and capacitors of 1 F. The normalized filter is obtained by setting

$$R_4 = \frac{2}{\dfrac{1}{250} - \dfrac{1}{10,000} - \dfrac{1}{10,000}} = 526.314 \; \Omega$$

To shift the center frequency to $f_0 = 5,000$ Hz we divide the capacitors by $\omega_0 = 2\pi 5,000$:

$$C_1 = C_2 = \frac{1}{2\pi 5,000} = 63.662 \; \mu F$$

Finally, to obtain practical values for all the components, we multiply all the resistors by 10^3 and divide all the capacitors by 10^3. The resistor and capacitor values become:

$$R_1 = R_2 = R_3 = R_5 = R_6 = 1 \; k\Omega$$
$$R_4 = 526.316 \; k\Omega$$
$$C_1 = C_2 = 0.06366 \; \mu F$$

The designer now specifies commercially available components and incorporates capacitor and/or resistor trimmers if necessary.

Note: The state-variable amplifier can also be used as a low-pass filter and as a high-pass filter if the proper output terminals are used as shown in Figure 6-5. This is an expensive version compared to other low-pass and high-pass filters, but has the advantage of low Q-sensitivity to circuit parameter variations. The DC gain and cutoff frequencies for the low-pass filter are

$$\text{Gain} = \frac{1 + R_5/R_6}{1 + R_3/R_4}$$

and

$$f_H = \frac{1}{2\pi} \left(\frac{R_6}{R_1 R_2 R_5 C_1 C_2} \right)^{1/2}$$

The high-frequency gain and cutoff frequency for the high-pass filter are

$$\text{Gain} = \frac{1 + R_6/R_5}{1 + R_3/R_4}$$

and

$$f_L = \frac{1}{2\pi} \left(\frac{R_4}{R_1 R_2 R_5 C_1 C_2} \right)^{1/2}$$

These expressions simplify for the case $R_1 = R_2 = R_3 = R_5 = R_6$ and $C_1 = C_2$, which we used for the band-pass filter.

7

Effect
of Component Characteristics
on Filter Performance
and How to Avoid Pitfalls

7.1 INTRODUCTION

An ideal operational amplifier draws no bias current into the two input terminals, and it has infinite gain for all frequencies. A real operational amplifier *does* draw bias current, it has finite gain, and the gain decreases with increasing frequency. In very demanding designs these properties of real amplifiers must be incorporated into the filter design, as must be real properties of passive components.

7.2 BIAS CURRENT

To ensure that there is no output offset voltage resulting from bias currents, it is necessary to have the currents into the inverting and noninverting terminals flow through the same effective resistance values so that the same voltage drops exist at both terminals. A DC analysis must be performed to trace the paths of the bias currents. As an example, we examine Figure 3-1. The figure is reproduced with the addition of the bias compensation resistor R_3, in Figure 7-1.

Remembering that AC signal sources offer negligible impedance to DC currents and that the output impedance of operational amplifiers is negligibly small, we can assume, for a DC analysis, that terminals A and C are grounded. The resistance R_+ seen from the noninverting terminal of the operational amplifier to ground is the series combination of R_1 and R_2, or

$$R_+ = R_1 + R_2$$

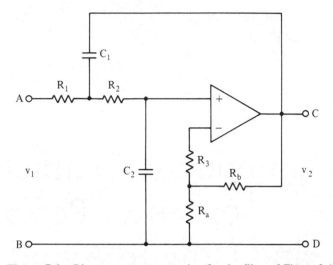

Figure 7-1 Bias current compensation for the filter of Figure 3-1.

The resistance R_- seen from the inverting terminal of the operational amplifier is equal to R_3 in series with the parallel combination of R_a and R_b. That is,

$$R_- = R_3 + R_a \| R_b$$

or

$$R_- = R_3 + \frac{R_a R_b}{R_a + R_b}$$

In order for the resistances seen from the noninverting and inverting terminals of the operational amplifier to be equal, we set $R_+ = R_-$, or

$$R_1 + R_2 = R_3 + \frac{R_a R_b}{R_a + R_b}$$

from which we solve for R_3,

$$R_3 = R_1 + R_2 - \frac{R_a R_b}{R_a + R_b}$$

Now, resistors R_a and R_b are selected to give the desired amplifier gain K, which together with the selection of R_1 and R_2 result in the desired filter characteristics as was explained with the help of tables and formulas in Chapter 3, and for similar configurations in other chapters of the book. To minimize the bias current effect on filter performance, a resistor R_3 computed from the equation given above is inserted in the circuit as shown in Figure 7-1.

7.3 AMPLIFIER GAIN

Representative amplifier gain versus frequency curves are shown in Figure 7-2. This figure shows curves for an internally compensated amplifier as well as for an uncompensated amplifier and for an externally compensated amplifier.

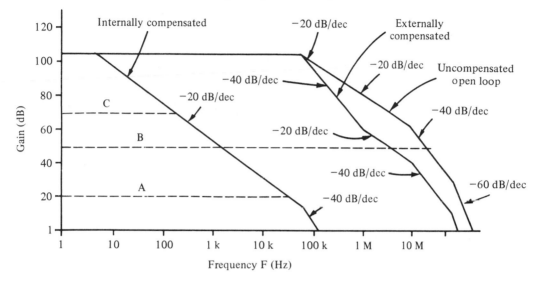

Figure 7-2   Amplifier compensation—uncompensated, externally compensated, and in ternally compensated gain-frequency response curve.

The flat portion of the gain-frequency response curve should extend at least to the frequency in the passband of the filter. In Chapter 3 we discussed low-gain amplifier filters and high-gain amplifier filters. From Figure 7-2 we conclude that an internally compensated amplifier is not, in general, usable in high-gain amplifier filters, since its break point (i.e., the 3-dB point) is at less than 10 Hz. But the amplifier is usable in low-gain amplifier filters. When an amplifier is used with negative feedback, as shown in Figure 7-3, then the closed-loop gain of the amplifier (i.e., the gain with the feedback resistors as shown) is given in terms of the ratio of the resistor values as

$$\text{Gain} = 1 + \frac{R_b}{R_a}$$

or in decibels

$$\text{Gain (dB)} = 20 \log_{10}\left(1 + \frac{R_b}{R_a}\right)$$

An amplifier is stable if its closed-loop frequency response (as given by lines A, B, and C for three different gains) intersects the open-loop gain response curve at a point where the latter has a slope of −20 dB/dec. Accordingly, for a gain of 10, with $R_b = 90$

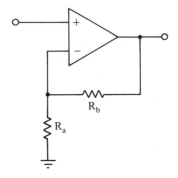

Figure 7-3   Setting amplifier gain.

kΩ and R_a = 10 kΩ, for example, or in decibels, for Gain (dB) = 20 $\log_{10} 10$ = 20 dB, the amplifier is stable and the response curve is flat up to approximately 20 kHz. The amplifier can be used in this configuration for filters up to about 10 kHz.

7.4 AMPLIFIER COMPENSATION

For higher-frequency filters, amplifiers that are not internally compensated must be used. Referring to Figure 7-2, we see that the uncompensated amplifier is stable with a gain corresponding to line C. This line (when extended) intersects the open-loop frequency response curve where the latter has a slope of −20 dB. The amplifier can be used in the 1-MHz frequency range. For a gain corresponding to line B the amplifier could be unstable, since the line intersects the open-loop frequency response curve where the latter has a slope of −40 dB/dec. To use the amplifier with this gain, it must be compensated by adding resistors and/or capacitors externally to modify the frequency response as shown in the figure. The steps that must be taken to compensate an amplifier are usually given by the manufacturer. With this external compensation, line B intersects the open-loop frequency response curve where the latter has a slope of −20 dB/dec.

In filters that use voltage followers (e.g., Figure 4-1) a compensation network can be incorporated, as shown in Figure 7-4. As the voltage frequency increases, more of the current fed back from the output through R_o flows through C_o and less into the amplifier, thus decreasing the negative feedback and maintaining higher gain. The component values are obtained from the relation

$$R_o C_o = \frac{1}{2.6 \times 2\pi f_a}$$

where f_a is the 0-dB frequency of the amplifier.

The Q of the high-Q band-pass filter of Chapter 7 is limited by the amplifier frequency response. To have a filter essentially independent of the break frequencies of the amplifiers, the amplifier must be selected so that

$$Q \ll A_o f_p / 4f_o$$

where f_p is the 3-dB break frequency of the amplifier, and f_o is the resonance frequency of the filter.

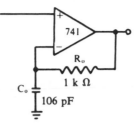

Figure 7-4 Frequency compensation to shunt feedback current at high frequency.

7.5 RESISTORS AND CAPACITORS

Nominal resistor values have been chosen and standardized by manufacturers to avoid waste in manufacturing. That is, *any* resistor manufactured subject to process control limitations falls within the tolerances of a nominal value. Accordingly, resistors with ±20% tolerance are available with the following nominal values: 10, 15, 22, 33, 47, 68, and 100 Ω and in multiples of 10 of these values; resistors with ±10% tolerance are available with the nominal values of 10, 12, 15, 18, 22, 27, 33, 39, 47, 56, 68, 82, and 100 Ω and multiples of 10 of these values; and ±5% tolerance resistors are available with the nominal values of 10, 11, 12, 13, 15, 16, 18, 20, 22, 24, 27, 30, 33, 36, 39, 43, 47, 51, 62, 68, 75, 82, 91, and 100 Ω and multiples of 10 of these values. Tighter tolerance resistors are often specified in fractions of ohms, and manufacturers' or suppliers' catalogs should be consulted.

The computed resistance and capacitance values are usually not of standard commercial products. When the filter is built with standard available components, trimmers must be incorporated in the design. Remember also that capacitor tolerances are on the order of ±20% and resistor tolerances are commonly ±10% or ±5% which would necessitate trimmers anyway. Components are subject to drifts with temperature. The filter must be tested over the required temperature range and adjusted or compensated as needed.

Metal-film resistors are available with long-term stability and with temperature coefficients as low as 1 ppm/°C. This makes them particularly suitable if resistor values must remain nearly constant over a wide temperature range. Least expensive are carbon-composition resistors. They can be used when cutoff frequencies with changes in temperature need not be tightly maintained. Wirewound resistors are not, in general, suitable for filters, because of their inductance. "Noninductive" *bifilar*-wound resistors are relatively expensive.

A large variety of capacitors is available. Ceramic capacitors and film capacitors are suitable for low-voltage electronic filters. The "temperature-stable" medium-K ceramic capacitors are a good compromise between temperature stability and dielectric constant value. Capacitors with zero-temperature coefficient of capacitance are available. These, however, have small dielectric constants and are limited in capacitance value to 5,000 pF. On the other hand, ceramic capacitors with higher dielectric constants are highly temperature- and frequency-dependent.

Popular in filter design are film capacitors. Polystyrene is an excellent dielectric for low-voltage applications and operating temperatures below 100°C. Polypropylene has similar characteristics in a somewhat smaller package. Polycarbonate and polyester (Mylar*) capacitors offer good performance over a wide temperature range. Teflon* can be used at temperatures up to 200°C.

The best capacitor for a given application depends on the required stability, temperature range, capacitance value, frequency, and other parameters. Detailed tables and curves describing the various commercial capacitor types can be obtained from manufacturers.

Thick-film and thin-film resistor arrays and capacitors are available. They should be considered for products involving a large number of filters. Film components can be trimmed to order during the fabrication process and can result in savings.

7.6 MORE INFORMATION

Detailed information about operational amplifiers and passive components is given in Parts II and III of this design guide. How to select the right amplifier and passive components based on their real-world properties is discussed and illustrated by examples in Part IV of this book.

*DuPont registered trademark.

Part II OPERATIONAL AMPLIFIERS

8

Understanding Operational Amplifier Characteristics

8.1 INTRODUCTION

The central element of modern linear electronic circuits is the operational amplifier. We distinguish between *voltage operational amplifiers* and *transconductance operational amplifiers*. A voltage operational amplifier is a high-voltage-gain, high-input-impedance, low-output-impedance amplifier. The term *operational* is used for historical reasons, as this type of amplifier was first developed to perform mathematical operations in analog computers. A transconductance operational amplifier is a high-input-impedance, high-output-impedance amplifier characterized by its linear transconductance, that is, by an output current proportional to the input voltage. The voltage operational amplifier is by far the most commonly used amplifier. In this book the term *operational amplifier* will imply ''voltage operational amplifier,'' unless qualified by the adjective *transconductance*. The operational amplifier underwent several stages in its development in rapid succession: vacuum tube amplifier, discrete transistor amplifier, and integrated circuit (IC) amplifier. The IC amplifier achieved a high degree of perfection, and a large variety of amplifiers is available.

8.2 SYMBOLS AND DEFINITIONS

The symbol of an operational amplifier is shown in Figure 8-1. All the voltages are referenced to ground, as shown in Figure 8-2. The voltages V^+ and V^- are the positive and negative power supplies, also referenced to ground. In most applications the operational amplifier is connected to positive and negative power supplies. The amplifier has no

47

Inverting input terminal

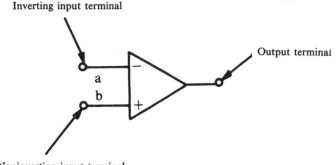

Output terminal

a

b

Noninverting input terminal

Figure 8-1   The operational amplifier symbol.

ground terminal, but the voltages in the amplifier circuit become referenced to ground through the power supplies and through the external input and output circuits. There are also circuits in which the operational amplifier operates from a single polarity power supply. In this case one terminal of the amplifier is connected to ground.

The operational amplifier can be utilized with three different configurations of the external circuitry. The signal voltage can be applied to the *inverting terminal* input (terminal a in Figure 8-2), to the *noninverting terminal* (terminal b), or as a differential input, between terminals a and b. Each configuration offers unique characteristics and applications. The manufacturer supplies the application engineer with a set of specifica-

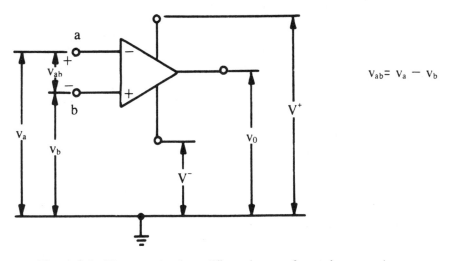

$$v_{ab} = v_a - v_b$$

Figure 8-2   The operational amplifier voltages referenced to ground. v_a = inverting input voltage. v_b = noninverting input voltage. v_o = op amp output voltage. V^+ = positive voltage of power supply. V^- = negative voltage of power supply.

tions for each operational amplifier type. These specifications impose constraints on the design and are ultimately responsible for particular choices of components. We shall explain the concepts and terms used in the specifications. In the last chapter of this part of the book we show, by example, how these specifications are included in the design procedures of practical amplifiers to clarify the concepts and definitions.

8.3 GAIN

The gain A of an operational amplifier is specified in terms of the *differential* input voltage v_{ab} and the output voltage:

$$A = -\frac{v_o}{v_{ab}}$$

The output voltage v_o is measured between the output terminal and ground. The negative sign in the equation indicates that if the voltage at terminal a is positive with respect to the voltage at terminal b, the voltage at the output terminal is *negative* with respect to ground. The gain A is a function of the input voltage frequency f: the gain is highest for DC and low-frequency signals, and decreases with increasing frequency. For stable operation the *gain versus frequency* characteristic must meet certain specifications, depending upon the

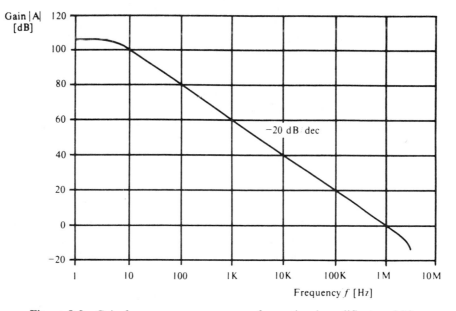

Figure 8-3   Gain-frequency response curve of operational amplifier type LM 741, an internally compensated amplifier. Note the low frequency of about 5 Hz at the "3-dB point" and the "bandwidth" of 1 MHz (i.e., frequency at unity gain or 0-dB gain).

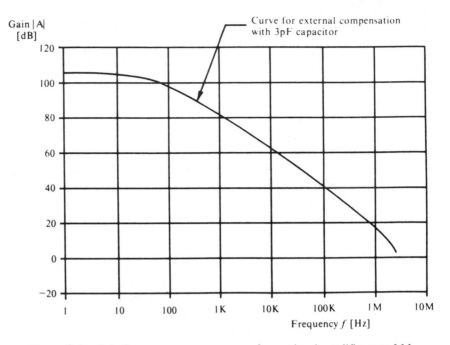

Figure 8-4 Gain-frequency response curve of operational amplifier type LM 748 with external compensation. Note the higher frequency of about 50 Hz at the "3-dB point" and the higher frequencies for comparable gains of the LM 741. For example, at 40 dB the LM 748 has a frequency response of 100 kHz compared to 10 kHz of the LM 741.

external circuitry in which the amplifier is used. To meet these characteristics, certain operational amplifiers are *compensated* internally by the manufacturer, while other operational amplifiers must be compensated by the user. Each type has a field of application. Chapter 10 deals in detail with the compensation problem.

The gain-frequency response curves of an internally compensated operational amplifier (type LM 741), and of an uncompensated operational amplifier (type LM 748) are shown respectively in Figure 8-3 and Figure 8-4. Note the difference in the magnitudes of the frequencies at the respective 3-dB points, and the differences in the slopes beyond the 3-dB points. Operational amplifiers are characterized by high DC gain values on the order of hundreds of thousands. The gain is often specified in decibels (dB). The decibel is defined as 20 times the logarithm (to the base 10) of a number:

$$\text{Gain (dB)} = 20 \log_{10} A$$

For example, a gain of 100,000 corresponds to a gain of 100 dB.

$$20 \log_{10} 100,000 = 20 \times 5 = 100 \text{ dB}$$

8.4 INPUT RESISTANCE

The input resistance of an amplifier is defined as the resistance between the inverting and the noninverting input terminals. It is shown symbolically as R_i in Figure 8-5. Values lie in the range from hundreds of kilohms to hundreds of megohms. The input of the LM 741 amplifier, for example, is specified as *"typical"* 1 MΩ and *minimum* 300 kΩ. The LM 741 has a junction transistor input stage. Input resistances of tens and hundreds of megohms are available in operational amplifiers with field-effect transistor input stages (e.g., the LH0042), or "super beta" transistor input stages (e.g., the LM108). The input capacitance is very small, on the order of a few picofarads.

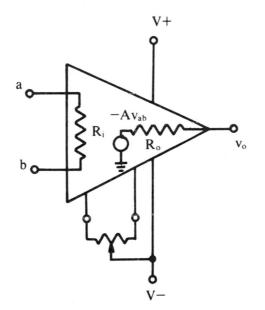

Figure 8-5   Voltage offset balancing circuit. (Also shown symbolically are input and output resistances.)

8.5 OUTPUT RESISTANCE

The output resistance of an operational amplifier constitutes circuitry responsible for the difference between the amplifier output voltage under no load and under load, due to current flowing in this part of the circuit under load. It is represented symbolically as R_o in Figure 8-5. It can be seen from the figure that if no load is connected to the output terminal of the amplifier, no current flows through R_o and the output voltage is equal to the no-load voltage: $v_{o(NL)} = -A\ v_{ab}$, where A is the gain of the amplifier and v_{ab} is the voltage between terminals a and b. A load connected to the amplifier draws a current i_L from the amplifier. The output voltage is reduced by the voltage $i_L R_o$ across R_o, giving an output voltage v_o, under load

$$v_o = v_{o(NL)} - i_L R_o$$

Output resistances of operational amplifiers are on the order of a few tens or hundreds of ohms. In an ordinary application of the operational amplifier, when peripheral feedback and input resistors are connected to the operational amplifier, the effective output resistance of the amplifier is only a few ohms or a fraction of an ohm.

8.6 COMMON-MODE GAIN (CMG) AND COMMON-MODE REJECTION RATIO (CMRR)

The operational amplifier ideally amplifies only the difference between the two signals applied to the inverting and noninverting input terminals. Ideally, if the two signals are equal in magnitude and in phase, they are not amplified at all, since their difference is zero. Such signals are called *common-mode* signals. They can be caused, for example, by 60-Hz signals induced from neighboring equipment or circuits, or by the drift of a differential output stage of a transducer feeding the operational amplifier.

In practical operational amplifiers, common-mode signals are amplified to some degree, since perfect matching of components inside the amplifier is impossible. This is called *common-mode gain (CMG)*. To minimize the effect of common-mode signals, operational amplifiers are designed to have low common-mode gain. The degree of interference seen at the output due to common-mode input depends on the relative magnitudes of the differential and common-mode input signals and on the *relative amplifications* of these signals. Accordingly, the *common-mode rejection ratio (CMRR)* is defined as the absolute value of the ratio of the differential gain divided by the common-mode gain:

$$CMRR = \left| \frac{A(w)}{CMG} \right| \tag{8-1}$$

As an example, suppose that $|CMG| = 3$ and $|A(w)| = 100,000$ for low frequencies. Then

$$CMRR = \frac{100,000}{3} = 33,333$$

and in decibels

$$CMRR \text{ (dB)} = 20 \log_{10} 33,333 = 90.46 \text{ dB}$$

8.6.1 Common-Mode Effect on the Output

The amplified common-mode input signal appears at the output together with the amplified differential input signal, the former being amplified by the common-mode gain,

CMG, and the latter by the differential gain $-A(w)$. With reference to Figure 8-2, the output v_o, is given by the relationship

$$v_o = -A(w)(v_a - v_b) + (CMG)\, v_{cm} \tag{8-2}$$

where v_a and v_b are respectively the voltages, measured with respect to ground, at terminals a and b, and v_{cm} is the common-mode voltage appearing at terminals a and b, as measured with respect to ground. Solving Equation 8-1 for *CMG* and substituting the result in Equation 8-2 we obtain

$$v_o = -A(w) \left[(v_a - v_b) \pm \left| \frac{v_{cm}}{CMRR} \right| \right] \tag{8-3}$$

which expresses the effect of the common-mode signal on the output in terms of the common-mode rejection ratio *CMRR* which is specified by the manufacturer. Note that in Equation 8-3 the positive value of $|v_{cm}/CMRR|$ is preceded by a $\pm$ symbol, as it is impossible to know *a priori* whether the common-mode signal is amplified by a positive or negative gain. The sign can be different for two ICs of the same type and model. It will be shown in Chapter 12 how common-mode signals affect the output of practical amplifier configurations.

8.6.2 Maximum Common-Mode Input

There is an amplitude limit to common-mode input voltages that may be applied to the inverting and noninverting terminals. Beyond this limit, the amplifier does not operate linearly. This limit is often given in the specification sheet for the amplifier. Maximum common-mode input voltage is an important consideration in applications such as comparators, where a reference signal is applied to one terminal and to the output switches when the signal at the other input terminal just exceeds the reference signal, or in applications where only a single voltage supply is used and the quiescent voltages at the input terminals are above ground.

8.7 VOLTAGE SUPPLY REJECTION RATIO (VSRR)

The characteristics of an operational amplifier are given for a certain set of test conditions including the voltage supplies such as ± 15 V. If the voltage supply drifts from its nominal value, the output voltage of the amplifier, in general, changes. The change in output voltage is expressed in terms of an equivalent differential input voltage which causes the same change in output voltage.

Thus, the effect of power-supply voltage variation, that is, the supply voltage rejection ratios, is given in terms of an equivalent input voltage per 1-V change in the power-supply voltage. The units are microvolts per volt or millivolts per volt. A typical value is 20 μV/V. The supply voltage rejection ratio is also given in decibels. For example,

$$20 \log_{10} 20 \ \mu V/V = 20 \log_{10} 20 \times 10^{-6} = -94 \ dB$$

Usually the absolute value, that is, 94 dB, is given in the specification.

Depending on the degree of output stability required, the designer must choose a suitable operational amplifier and a suitable power supply. Obviously, the poorer the amplifier's voltage supply rejection ratio, the better the regulation and drift characteristics of the power supply must be.

Example

Consider an amplifier of the configuration shown in Figure 8-6(a) with $R_1 = 10 \ k\Omega$ and $R_2 = 50 \ k\Omega$. The *VSRR* of the amplifier is specified as 200 $\mu V/V$. A voltage drop of 100 mV at the V^+ terminal is caused by other components of the system. By how much will the output voltage change as a result of the change in power-supply voltage?

Solution. The supply voltage drop of 100 mV, or 0.1 V, has the same effect on the output as the effect caused by 20 μV, or 20×10^{-6} V, between the input terminals of the operational amplifier. The effect is calculated by multiplying this voltage by the quantity $(1 + R_2/R_1)$.

$$v_o = \left(1 + \frac{50 \times 10^3}{10 \times 10^3}\right) 20 \times 10^{-6} = 6 \times 20 \times 10^{-6} = 120 \times 10^{-6} \ V$$

or 0.120 mV.

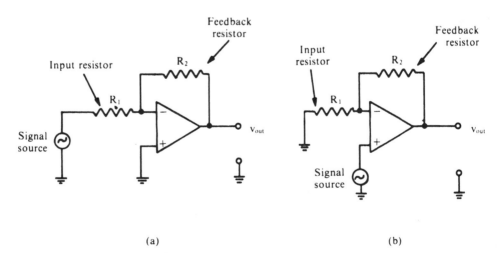

(a)   (b)

Figure 8-6   Basic operational amplifier applications. (a) The inverting amplifier. (b) The noninverting amplifier.

8.8 INPUT OFFSET VOLTAGE

An ideal amplifier has zero output voltage for zero input voltage. Since real components cannot be perfectly matched, and since components in symmetrical positions may be at slightly different temperatures, the differential amplifier stages of an operational amplifier are, in general, not balanced perfectly. Consequently, there is a finite output voltage with zero input voltage. The *offset voltage* is defined as the differential voltage that must be applied to the *input terminals* to reduce the output voltage to zero. This voltage can be a few millivolts in magnitude. The polarity of the offset voltage is not known *a priori*. The *quiescent output voltage* resulting from the offset voltage depends on the closed-loop amplifier design and is approximately equal to the offset voltage multiplied by the closed-loop gain of the amplifier.

To balance the amplifier, a potentiometer is connected to a pair of balancing terminals, as shown in Figure 8-5. The potentiometer is adjusted to obtain zero output voltage. A resistance value for the potentiometer is usually supplied by the manufacturer; it is typically 10 kΩ or 50 kΩ.

Since the offset voltage, like other characteristics of the amplifier, is temperature-sensitive, the offset voltage can be balanced out only at a given temperature. It is therefore necessary to set the balancing potentiometer at the operating temperature and to check the output over the relevant temperature range.

8.9 INPUT BIAS CURRENT

DC currents must be applied to both input terminals of an operational amplifier to bias the amplifier transistors properly. The circuitry that allows this bias current to flow is supplied externally to the operational amplifier and is the responsibility of the application engineer. The manufacturer specifies the magnitude of the bias current for the operational amplifier. In practical operational amplifiers, the two input transistors are not matched perfectly, and in general the two bias currents I_{b1} and I_{b2} are different. The specified value I_b is the average of the two:

$$I_b = \frac{I_{b1} + I_{b2}}{2}$$

Improper design of the bias circuit causes an undesirable output voltage which may exceed that caused by the input offset voltage.

To understand problems that can arise from improper bias circuitry and to learn how to alleviate such potential problems, we must pause for a moment and look at the two basic applications of the operational amplifier shown in Figure 8-6. The inverting amplifier has an output voltage that is 180 deg. out-of-phase with the input signal voltage, while the noninverting amplifier has an output voltage in phase with the input signal voltage. Each configuration offers certain unique features that are described in Chapter 11, Sections 11.1 and 11.2. To consider the bias problem, we recall that to study DC characteristics of an electrical circuit, AC voltage sources are replaced with short circuits. Shorting the signal sources in Figure 8-6, both amplifier configurations take the form shown in Figure 8-7(a).

Because of the high gain of operational amplifiers and because of the negative feedback in the amplifier of Figure 8-7(a), there is a tendency of the potentials at the inverting input terminal a and the noninverting terminal b to be the same. If, say, the potential at a were higher than the potential at b, then there would be a negative output voltage. The output voltage is fed back through R_2, lowering the potential at a. If the open-loop gain of the operational amplifier were infinite, the difference between the potentials at a and b would be zero. Since terminal b is held at ground potential, we say that terminal a is at virtual ground. In first-order analyses, we can assume $v_a - v_b = 0$, where v_a and v_b are respectively the potentials (or voltages with respect to ground) at terminals a and b.

We now study the bias problem. Since terminal a is at virtual ground potential, the bias current into terminal a must flow through R_2. (If bias current were to flow through R_1, there would be a voltage drop across R_1 causing the potential at a to be below ground, which contradicts the imposed condition that a is at ground potential.) To allow the bias current I_b to flow through R_2, the output voltage rises to a value of $I_b R_2$, equal to the voltage drop across R_2. The noninverting terminal b receives bias current from ground through the base emitter junction, and the current flows to the negative supply terminal. We thus find that the circuit of Figure 8-7(a) has an output voltage $I_b R_2$ with no signal input. This is an undesirable condition, since we want zero output with zero input. This undesirable condition is avoided by the addition of a compensation resistor R_3, as shown in Figure 8-7(b). The value of R_3 must be equal to the value of the parallel combination of R_1 and R_2.

$$R_3 = R_1 \| R_2 = \frac{R_1 R_2}{R_1 + R_2}$$

With R_3 in the circuit, the bias current into terminal b flows through R_3 producing a voltage at b of magnitude

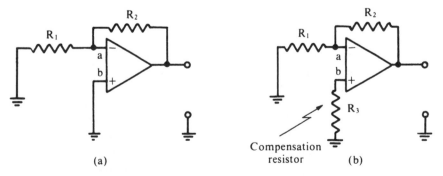

(a)   Compensation resistor   (b)

Figure 8-7   The amplifier of Figure 8-6 with short-circuited voltage sources. (a) Not compensated for bias imbalance. (b) Compensated for bias imbalance.

$$v_b = I_b R_3$$

Since as noted above $v_a - v_b = 0$, the voltage at terminal a is

$$v_a = -I_b R_3$$

This value in turn determines the currents through R_1 and R_2. Let $I_b{'}$ be the current through R_1, and $I_b{''}$ the current through R_2. We have

$$I_b{'} = -\frac{v_a}{R_1} = \frac{I_b R_3}{R_1}$$

and, for zero output voltage,

$$I_b{'''} = \frac{v_a}{R_2} = -\frac{I_b R_3}{R_2}$$

The total bias current into terminal a is

$$I_b{'} + I_b{''} = I_b R_3 \left(\frac{1}{R_1} - \frac{1}{R_2} \right) = I_b R_3 \frac{R_1 + R_2}{R_1 R_2}$$

Since we chose $R_3 = \frac{R_1 R_2}{R_1 + R_2}$, therefore $R_3 \frac{R_1 + R_2}{R_1 R_2} = 1$ and

$$I_b{'} + I_b{'} = I_b$$

Thus, we get the required bias current while maintaining zero output voltage as required. Thus, the choice of R_3.

Example

Let $I_b = 200$ nA, $R_1 = 2$ kΩ, and $R_2 = 50$ kΩ. Then $R_3 = R_1 \| R_2 = 1.9$ kΩ, $v_a = v_b = -0.38$ mV, and $v_o = 0$ V. If R_3 were not used, we would have $v_a = v_b = 0$ and $v_o = I_b R_2 = 10$ mV.

The lowest permissible values of R_1, R_2, and R_3 are determined by input impedance and gain requirements as is made clear in Chapter 12 in which design examples are presented; the upper limit is determined by the input offset current discussed in the next section.

8.10 INPUT OFFSET CURRENT

In real operational amplifiers the bias currents I_{ba} and I_{bb} into terminal a and b, respectively, are not equal. The difference $|I_{ba} - I_{bb}|$ is the offset current. The offset current causes an output of the amplifier in the absence of an input. Consider Figure 8-7(b). Let $R_2 = 500$ kΩ and $R_1 = 100$ kΩ. We compute $R_3 = R_1 \| R_2$ giving $R_3 = 83.3$ kΩ. Let $I_{bb} = 300$ nA and $I_{ba} = 400$ nA. Then $v_b = -I_{bb} R_3 = -2$ mV. But $v_a = v_b$, giving the current through R_1 as $I_b' = -v_b/R_1 = 250$ nA. The difference I_b'' between I_{ba} and I_b' flows through R_2. Thus $I_b'' = I_b - I_b' = 150$ nA and $v_o = v_a - R_2I_b'' = 10$ mV. We see that the input offset current causes an undesirable output just as does the input offset voltage. Usually, the application engineer designs the circuit so that the effect of the offset current on the output is less than the effect of the offset voltage. The offset current puts an upper limit on resistor values that can be used to keep the output voltage within the prescribed specifications.

8.11 GAIN-BANDWIDTH PRODUCT

The gain-bandwidth product of an operational amplifier is given by the product of the DC gain of the amplifier and the 3-dB frequency. The 3-dB frequency is the frequency at which the gain is down by 3 dB from the DC gain. Since the manufacturer does not know the closed-loop gain that will be employed by the user, the open-loop gain-bandwidth product is specified. This is a meaningful figure if it is known that the gain rolloff beyond the 3-dB point is −20 dB/dec (−6 dB/oct), as in Figure 8-3. Knowing this, it is possible to predict the gain-bandwidth product for any closed-loop gain. The intersection of the open-loop transfer curve and a straight line drawn parallel to the abscissa at a height equal to the closed-loop gain gives the 3-dB frequency. Multiplication of the closed-loop gain by this frequency gives the gain-bandwidth product. *Note:* For a rolloff of −20 dB/dec, the open-loop response curve intersects the 0-dB (unity gain) line at a frequency equal to the gain-bandwidth product of the amplifier (1 MHz in the example of Figure 8-3).

When the open-loop response of an amplifier rolls off at a greater slope than 20 dB/dec, the open-loop gain-bandwidth product has no significance for closed-loop operation. In this case, the manufacturer usually specifies the unity gain frequency, which is somewhat more indicative of closed-loop capabilities, but in either case the usefulness of this information is limited and the complete frequency response curve is needed for closed-loop performance prediction. It is also important to keep in mind that the response curve is changed entirely if the amplifier is *compensated* for stability. Compensation is discussed in Chapter 10.

8.12 SLEW RATE

The *slew rate, SR,* is defined as the highest possible rate of change for the amplifier output voltage.

$$SR = \left(\frac{dv_o}{dt} \right)_{max}$$

This specification is important for *large signal* operation. The slew rate of an amplifier is limited by the currents available for charging parasitic capacitances or deliberately introduced capacitors in the amplifier. At certain points in the circuit of an operational amplifier the response to a signal at the input terminals is dependent on a *current source*. This current charges the various capacitances building up the output voltage in response to the input voltage. The voltage v_o across a capacitor is equal to the quotient of the charge q and the capacitance C:

$$v_c = q/C$$

For a constant charging current I_o, this equation becomes

$$v_c = \frac{I_o}{C} t$$

The rate of voltage change across the capacitor is obtained from this equation as

$$\frac{dv_c}{dt} = \frac{I_o}{C}$$

The slew rate is given by

$$SR = \frac{I_{max}}{C}$$

where I_{max} is the maximum current available in the circuit of the operational amplifier to charge the capacitors.

The slew rate imposes a frequency as well as an amplitude limitation. An amplifier may respond faithfully to an input signal of a certain frequency, but may give a distorted output in response to an input signal of the *same* frequency when the output voltage called for is of a higher amplitude. A voltage of a given frequency and amplitude undergoes a higher rate of change than a voltage of the same frequency but of a lower amplitude, as is illustrated in Figure 8-8. The requirement for a large output amplitude can be caused by a larger input signal and/or by a higher closed-loop gain of the amplifier. By similar reasoning it is also easy to see that of two signals of the same amplitude, the one of the higher frequency requires a larger rate of change.

The amplitude and frequency of a signal are related to the required slew rate (SR). Consider an output voltage of amplitude V_o and frequency f (angular frequency $\omega = 2\pi f$) given by

$$v = V_o \sin\omega t$$

The rate of change is

$$\frac{dv}{dt} = \omega V_o \cos\omega t$$

and the maximum rate of change is given when $\cos\omega t = 1$:

$$\left(\frac{dv}{dt} \right)_{max} = 2\pi f V_o$$

Hence, an undistorted sine wave, of amplitude V_o and frequency f at the output, requires a slew rate, SR,

$$SR \geq 2\pi f V_o$$

For example, if an output of 7 V peak at a frequency of 10 kHz is required, an amplifier with a slew rate of at least 0.44 V/μs must be used.

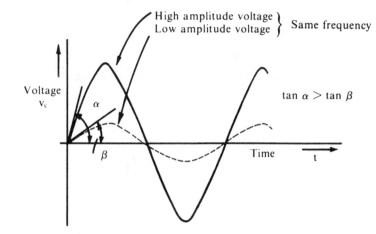

Figure 8-8 Rate of change of voltage as dependent upon amplitude.

8.13 CROSSOVER DISTORTION

In the design of equipment using operational amplifiers it is usually assumed that the gain of the amplifier is linear over the operating range. This is not true for small signals near the zero crossover *if* the amplifier uses a *class B* output stage. Two requirements for an output stage are low output impedance and high-power capabilities. An emitter-follower output stage possesses these properties, but has the drawback of high-power dissipation in the emitter resistor, which causes heating of the amplifier chip and low efficiency of the amplifier. These shortcomings are removed if the output stage is operated *class B* using a complementary pair (npn and pnp) of transistors, but different shortcomings are introduced.

The zero crossover when one transistor turns on and the other transistor turns off is never perfect. The input-output characteristic of the amplifier is then as shown in Figure 8-9. The output is not only distorted, but also the open-loop gain (which is given by the slope of the curve) is considerably smaller for small signals than for larger signals. Since the zero crossing region is relatively narrow, it is usually of no concern. But in applications where this degree of distortion or the small gain effect cannot be tolerated, the designer must avoid the use of an operational amplifier with a *class B* output stage.

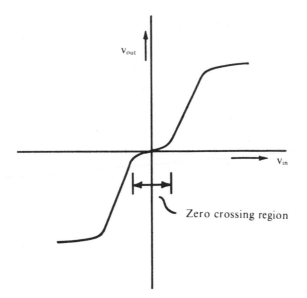

Figure 8-9   Input-output characteristic of *class B* amplifier.

8.14 RATED OUTPUT

The manufacturer usually specifies maximum output voltage and maximum output current. The values represent limits above which the output is distorted, or the amplifier destroyed. If the voltage is exceeded, the amplifier enters saturation. This results in distortion, and there is a lengthy recovery time. It should be noted that some amplifiers—the comparators—are used in saturated operation and are especially designed for relatively short recovery times.

Many operational amplifiers are built with current overload protection. The circuit operates in the normal mode for load currents below a specified point, and enters into the overload mode when the rated current is exceeded, resulting in distortion. The user must be aware that lack of overheating does not imply normal operation.

8.15 POWER DISSIPATION

Two *power dissipation* values can be given: *device dissipation*, P_D, and *total dissipation*, P_T. The device dissipation is the power that can be safely consumed by the amplifier. The total dissipation refers to the amplifier plus load dissipation. Thus, permissible load power $P_L = P_T - P_D$. Any two values may be given by the manufacturer. The load power may be given in terms of maximum load voltage and maximum load current for resistive loads. The product of voltage and current gives power.

8.16 INPUT OVERLOAD PROTECTION

Some amplifiers have an input overload protection. For differential input voltages below a specified value, the protection circuit is inactive. For larger differential input voltages, any excess current that would damage the input transistor is redirected to bypass the transistor, resulting in distorted output. The designer must be careful to keep the signal values below levels that activate the protection circuit to maintain distortionless operation. If an application requires large differential input voltages, the designer must use an amplifier without input overload protection.

8.17 SUPPLY CURRENT DRAIN

Since normally more than one IC is supplied from one power supply, care must be taken that the total current drain does not exceed the power-supply rating. Note that some amplifiers can be operated from a wide range of supply voltages and that the current drain may depend on the voltage. In other amplifiers, the current drain may be independent of the supply voltage.

8.18 TRANSIENT RESPONSE

When a pulse is applied to the input of the amplifier, the output responds as shown in Figure 8-10. The manufacturer may specify the percentage *overshoot, rise time* (10% to

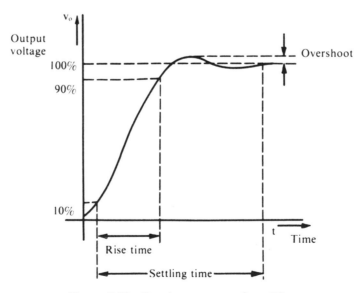

Figure 8-10   Transient response of amplifier.

90% of final output), and *settling time,* as well as the slew rate discussed in Section 8-12. All these specifications are defined in the figure.

8.19 INPUT CAPACITANCE

The parasitic input capacitance is usually in the range of 1 to 10 pF and is specified by the manufacturer.

8.20 AMPLIFIER NOISE

Amplifier noise is an important consideration in low-level signal application. The complete design must consider noise originating from the amplifier as well as other circuit components. The noise can be expressed in terms of input voltage or input current that causes the same output voltage as that caused by the noise. Curves expressing the amplifier noise *referred to the input,* are supplied by the manufacturer. See, for example, the noise characteristics for the LM 741* operational amplifier shown in Appendix A1.

8.21 TEMPERATURE, FREQUENCY, SUPPLY-VOLTAGE, SOURCE-RESISTANCE, AND LOAD-RESISTANCE DEPENDENCIES OF AMPLIFIER CHARACTERISTICS

The characteristics of an operational amplifier are specified under certain *operating* or *test* conditions. These conditions may include load resistance, temperature, and other parameters. So that the reader can study orders of magnitudes, units, and parameter dependencies of the various characteristics, a complete set of specifications of the bipolar LM 741** and the FET input LH0042 amplifiers is included in Appendix A3.

---

*The LM 741 characteristics are given by the LM 747 data sheets. The LM 747 is a package containing two LM 741s. The characteristics are identical.

**The LM 747 shown in the appendix is a package containing two LM 741 opamps. The characteristics shown are identical with those of the LM 741.

9

Dual- and Single-Power-Supply Amplifiers, Low-Voltage Amplifiers, and Transconductance Amplifiers

9.1 INTRODUCTION

Most operational amplifiers are designed for use with dual-voltage power supplies (± 15 V or ± 12 V are typical). This permits input and output signals to swing above and below ground voltage without the need for coupling capacitors. However, a dual-voltage power supply can be more expensive than a comparable single-voltage power supply of twice the voltage. Amplifiers usually supplied by a dual-voltage power supply can be operated from a single-voltage power supply, in which case an external voltage divider must be added to set the input halfway between the supply-voltage extremes.

9.2 SINGLE-VOLTAGE-POWER-SUPPLY AMPLIFIERS

The CA3015 (RCA) amplifier normally operates from a ± 12-V supply. It can also be operated from a 24-V supply. A diagram showing the power-supply connections is shown in Figure 9-1(a). *Note:* The input and output are capacitively coupled to the amplifier.

With the increased interest in applications of ICs in the automotive industry, operational amplifiers are available that are explicitly designed for use with a single-voltage power supply. Some can also be used with a dual-voltage power supply. For example, the LM 2902 quad operational amplifier (National Semiconductor) contains four operational amplifiers in one package. These amplifiers can be operated from a wide range of power supplies from 3 V to 26 V, or from ± 1.5 V to ± 13 V. Thus, the amplifier can be operated from a 5-V power supply in a digital system, eliminating the need for a separate dual-voltage power supply. This amplifier can be used to sense signals near ground. An

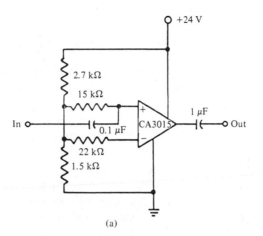

(a)

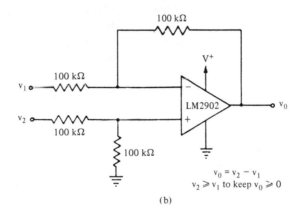

$$v_0 = v_2 - v_1$$
$$v_2 \geqslant v_1 \text{ to keep } v_0 \geqslant 0$$

(b)

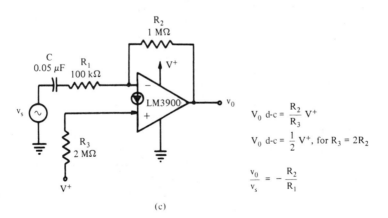

$$V_0 \text{ d-c} = \frac{R_2}{R_3} V^+$$

$$V_0 \text{ d-c} = \frac{1}{2} V^+, \text{ for } R_3 = 2R_2$$

$$\frac{v_0}{v_s} = -\frac{R_2}{R_1}$$

(c)

Figure 9-1 Amplifier operated from single-voltage power supply. (a) CA 3015 (b) LM 2902 (c) LM 3900.

example showing the application of the device in a difference amplifier is shown in Figure 9-1(b).

A popular single-polarity power-supply amplifier is the LM 3900 quad operational amplifier (National Semiconductor). This amplifier is different from the conventional amplifiers in that the input stage is not a voltage differential amplifier, but a current differential amplifier. The inverting input stage is a common emitter stage. The noninverting input stage is provided by a circuit known as a *current mirror* circuit. This results in an input stage in which currents are compared or differenced. It is called a *Norton* differential amplifier. There is no limit to the common-mode input voltage range. This is useful in high-voltage comparator applications. By making use of input resistors to convert input voltages to input currents, all of the standard operational amplifier applications can be realized. An application example is shown in Figure 9-1(c), where we show an inverting amplifier. Note that the DC level at the output is determined by a resistance ratio. In particular, the DC level can be set halfway between the power supply voltage and ground by choosing $R_3 = 2R_2$. An important practical advantage of this amplifier is that no voltage-dividing network is needed to establish the bias level of the input stage, which would reduce the input impedance. The noninverting input is simply connected to the power supply through a resistor. Note that the Norton amplifier is identified by two arrows, one between the inverting and noninverting terminals, and one into the noninverting terminal.

9.3 LOW-VOLTAGE AMPLIFIERS

As demands for miniature, portable, battery-operated instruments increase, the supply of low-voltage amplifiers becomes more pressing. For portable radios and calculators, 9-V batteries are quite adequate, but for miniature devices such as watches and hearing aids, the conventional 9-V battery is much too large. Miniature batteries the size of an aspirin tablet or smaller, of about 1.4 V, are now available. Another demand for low-voltage battery-operated equipment is low current drain.

The LM 4250 (National Semiconductor) opamp can be operated from voltages as low as 2 V or ± 1 V. This is clearly satisfactory for operation with two batteries. An external bias current resistor controls the input bias current, input offset current, quiescent power consumption, slew rate, input noise, and the gain-bandwidth product. The standby power consumption can be adjusted to as low as 500 nW.

For still lower voltages, the ICL 7641 (Intersil) opamp has been introduced. It can operate from voltages as low as 1 V or ± 0.5 V and is thus suitable to operate from a single battery. It has an input impedance of 10^{12} Ω and a typical input current of 1 pA. The opamp is a CMOS device and consequently the output swing ranges within a few milli-volts of the supply voltage. The standby quiescent current is 1 mA, the unity gain bandwidth is 1 MHz, and the slew rate is 1.6 V/μs. The 7641 package contains four opamps. Other versions of the same amplifier with fewer opamps per package are also available. In these, the extra available package pins are used to connect external resistors to the amplifiers, one per amplifier, which allows the programming of the amplifier characteristics to operate at a low standby current of 10 μA.

9.4 LOW-NOISE AMPLIFIERS

In many cases the signal that needs to be filtered is very low and it is necessary to amplify it before filtering. It is important to have a large signal-to-noise ratio, adding as little noise as possible by the amplification stage. There are so called "low-noise" amplifiers (see Appendix A8). For proper design the *right* low-noise amplifier must be chosen. This requires the selection of the amplifier with the best noise characteristics relative to the signal source output resistance.

Some manufacturers provide noise figure vs. source resistance curves such as shown in Figure 9-2. The noise figure is given as a function of the output resistance of the signal source connected to input terminals of the amplifier. One now compares the noise figures for the correct frequency and source resistance of several operational amplifiers and chooses the one with the lowest noise figure for these particular parameters.

If the noise figure (NF) curves are not given, the best amplifier can be determined from the noise voltage and noise current curves as follows. For the desired frequency, the noise figure is calculated using the formula given below for the particular source resistance.

$$NF = \frac{(\bar{e}_{ni}^2 + \bar{i}_{ni}^2 R_s) + 4KTR_s}{4kTR_s}$$

where $\bar{e}_{ni}^2$ and $\bar{i}_{ni}^2$ are respectively the noise voltage (in volts) and current (in amperes) and R_s is the source resistance (in ohms); T is the absolute temperature in degrees Kelvin, and k is Boltzmann's constant ($k=1.33 \times 10^{-23}\ J/°K$).

The amplifier that gives the lowest noise figure for the particular source resistance will introduce least noise.

9.5 WIDEBAND AMPLIFIER

Operational amplifiers are in general low-frequency devices. Typically, unity gain (i.e., 0 dB gain) occurs at about 1 MHz for internally compensated amplifiers and 4 or 5 MHz for uncompensated amplifiers. The 3 dB breakpoint is about 10 Hz. When filtering must be done at higher frequencies, wideband amplifiers must be used.

Some manufacturers supply special wideband amplifiers. (See Appendix A9.) The RCA CA3130 opamp, for example, has an uncompensated unity gain crossover frequency of 15 MHz; with a 47-pF compensation capacitor, the unity gain crossover frequency is 4 MHz. It is a BiMOS operational amplifier that combines the advantages of both CMOS and bipolar transistors on a monolithic chip.

The RCA CA3100 opamp has an open-loop gain of 42 dB at 1 MHz and a unity-

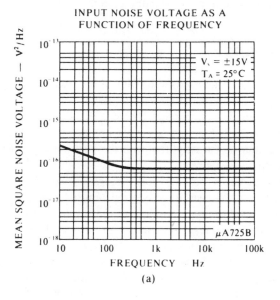

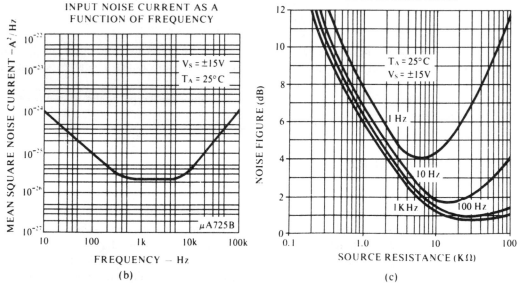

Figure 9-2 Typical input noise characteristics for the SSS 725.

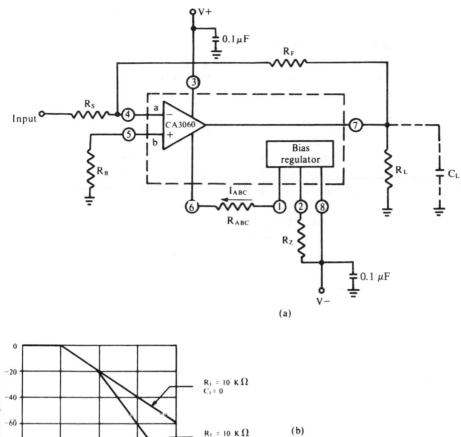

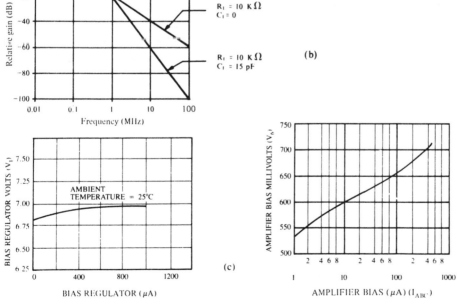

Figure 9-3 Inverting transconductance amplifier. (a) Schematic. (Reprinted by permission of RCA Solid State Division, now Harris Semiconductor) (b) Effect of capacitive loading on frequency response. (c) Amplifier bias voltage and bias regulator voltage as function of current.

gain crossover frequency of 38 MHz. The 3-dB corner frequency is 110 kHz. The RCA CA3023 has a unity-gain crossover frequency of about 50 MHz and a 3-dB corner frequency of 16 MHz. The voltage gain at 5 MHz is 53 dB.

9.6 TRANSCONDUCTANCE OPERATIONAL AMPLIFIERS

Whereas the basic characteristic of the voltage operational amplifier, discussed so far, is the forward voltage gain, the basic characteristic of the transconductance amplifier is the *transconductance*, g_m: that is, the ratio of the amplifier output current i_{out} to the input voltage v_{in}: $g_m = i_{out}/v_{in}$. The output of the amplifier is a current proportional to the input voltage. This is accomplished by providing a very large output impedance in contrast to a very low output impedance in voltage amplifiers. The transconductance amplifier is a high-impedance circuit; it is meant for low currents (microamps) and low dissipation of a few milliwatts, or a fraction of a milliwatt and as little as a few nanowatts. Of particular interest is the provision that allows the designer to adjust the transconductance of the amplifier. This is accomplished by providing the appropriate amplifier bias current I_{ABC}. In practice, this is accomplished simply by the selection of the value of a bias resistor R_{ABC}. An inverting amplifier circuit that uses a transconductance amplifier is shown in Figure 9-3(a). The current through R_L is proportional to the voltage applied to the input terminal. The design procedure for the amplifier is given in Chapter 12, Section 12.3.

The characteristics of the IC, including *input offset voltage, input bias current, peak output current, peak output voltage, forward transconductance,* and *output resistance* are determined by the amplifier bias current. A set of curves that give the characteristics of the CA 3060 (RCA) is given in Appendix A4.

10
Amplifier Stability

10.1 INTRODUCTION

Improper application of an operational amplifier can result in instability (oscillations). Stability can be ensured if certain rules are followed regarding phase compensation. In this chapter we discuss amplifiers that are internally compensated (by the manufacturer), and amplifiers that must be compensated externally by the application engineer.

10.2 PHASE COMPENSATION

Amplifier instabilities and oscillations occur if a voltage is fed back from the output to the input with a total loop phase shift of 360 deg. and unity gain. Oscillations can be initiated by noise anywhere in the circuit. The right conditions for oscillations usually occur between 10 kHz and 30 MHz. To prevent such oscillations from occurring, the amplifier is *compensated* by means of RC networks. These networks modify the open-loop frequency response of the amplifier, which in turn results in stable closed-loop operation. The simplest compensation method consists of reducing the overall gain of the amplifier with resistive networks; this, of course, defeats the main desirable characteristic of operational amplifiers, high open-loop gain. A practical method consists of reducing the gain in certain frequency ranges. This is always associated with a change in the phase-shift-frequency response of the amplifier. In fact, stability criteria have evolved about phase shifts and *phase margins*. Since, however, the amplitude-frequency and phase-shift-frequency characteristics are intrinsically related (both being dependent on the same resistive and capacitive components), the necessary information for our purposes can be obtained directly from the amplitude-frequency characteristic.

Stability is assured if the *closed-loop* gain transfer function meets the *open-loop* gain transfer function where the open-loop transfer function has a slope of -20 dB/dec. Figure 10-1 shows the open-loop transfer curve of an amplifier and closed-loop transfer curves for two different feedback ratios. Curve "a" corresponds to a stable amplifier, while curve "b" corresponds to an unstable amplifier. It is seen that the stable amplifier incorporates less feedback (i.e., has higher closed-loop gain) than the unstable amplifier. Given the open-loop response curve of an amplifier, the designer can determine the closed-loop gain range for which the amplifier is stable. The bandwidth of the stable amplifier is given by f_B, the frequency at the 3-dB attenuation point. The lower the closed-loop gain, the wider the bandwidth.

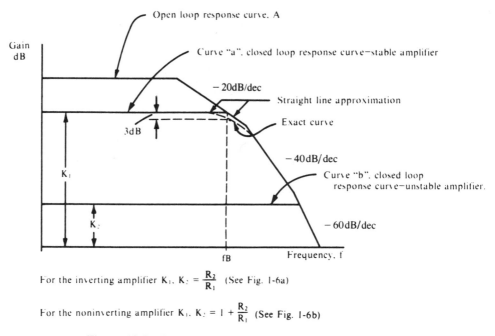

For the inverting amplifier K_1, $K_2 = \dfrac{R_2}{R_1}$ (See Fig. 1-6a)

For the noninverting amplifier K_1, $K_2 = 1 + \dfrac{R_2}{R_1}$ (See Fig. 1-6b)

Figure 10-1   Response curves of stable and unstable amplifier.

10.3 INTERNALLY COMPENSATED AMPLIFIERS

The obvious advantage of amplifiers that are internally compensated by the manufacturer is the convenience given to the application engineer not to have to compensate the amplifier. The disadvantage is that these amplifiers have very low cutoff frequencies. Since it is the intention to have an internally compensated amplifier stable under all probable applications, the open loop is modified to have a -20-dB/dec slope throughout the frequency range, starting from the break frequency of about 5 Hz as shown by the internally compensated line in Figure 10-2. Lines A, B, and C in the figure show, for example, that the amplifier is stable for gains of 20, 50, and 70 dB, since for each of these gains the closed-loop curve meets the open-loop-gain curve where the latter has a −20-dB/dec slope. Note that the 3-dB frequencies are at 60 kHz, 2 kHz, and 150 Hz, respectively.

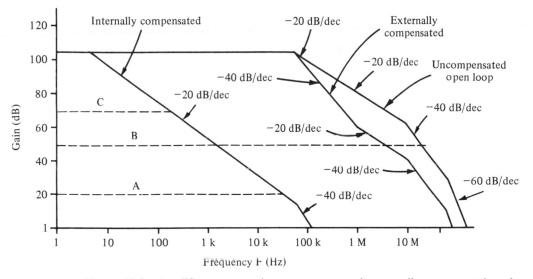

Figure 10-2 Amplifier compensation—uncompensated, externally compensated, and internally compensated gain-frequency response curve.

10.4 UNCOMPENSATED AMPLIFIERS

Uncompensated amplifiers are more versatile, since they can be compensated to give closed-loop gains with a wider frequency response than an internally compensated amplifier designed for the same closed-loop gain. The compensation responsibility lies, however, with the application engineer.

Consider the uncompensated gain-frequency curve shown in Figure 10.2. If the uncompensated amplifier is used for a closed-loop gain of 50 dB, the closed-loop gain-frequency curve meets the open-loop curve where the latter has a slope of −40 dB/dec and the amplifier is likely to be unstable. If the open-loop curve is reshaped as shown by the "externally compensated" curve, the amplifier is stable with a closed-loop gain of 50 dB because the closed-loop curve meets the compensated open-loop curve at a −20-dB/dec slope of the latter. The amplifier has an 8-MHz 3-dB point as compared to the 2-kHz 3-dB point of the internally compensated amplifier. Note that the externally compensated amplifier is likely to be unstable for a closed-loop gain of 20 dB as well as 70 dB. This example illustrates very clearly how the externally compensated amplifier can be

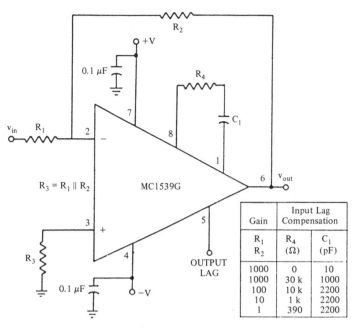

(a)

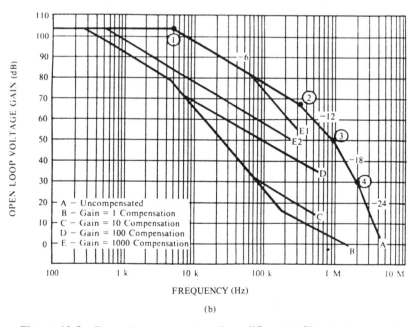

FREQUENCY (Hz)

(b)

Figure 10-3 External compensation of amplifier. (a) Circuit diagram. (b) Response curves. (Reprinted with permission of Motorola Semiconductor Products, Inc.)

custom-tailored for a particular application, while the internally compensated amplifier cannot.

The uncompensated curve can be reshaped to that of the externally compensated curve by connecting an appropriate network to the "compensation" terminals of the amplifier. The required procedure for compensation for practical amplifiers is supplied by the manufacturer. For example, Figure 10-3 gives compensation instructions in the form of diagrams, tables, and response curves for the MC1359 operational amplifier (Motorola). The information gives the required values of the compensating network components R_4 and C_1 for particular values of desired closed-loop gain. The two 0.1-μF capacitors shown in the figure are power-supply decoupling capacitors removing variations in the power supply caused by other devices in the system. It is good practice to use such decoupling capacitors for all ICs in a system.

11

Amplifier Configurations

11.1 INTRODUCTION

In this chapter, we introduce the three basic amplifier configurations for DC and AC operation, their characteristics, and their respective merits. The configurations are for the *inverting, noninverting,* and *differential* amplifiers.

11.2 THE INVERTING AMPLIFIER

The inverting amplifier configuration is shown in Figure 11-1. Resistors R_1 and R_2 determine the voltage gain of the amplifier; R_1 determines also the input resistance, and R_3, set equal to the parallel combination of R_1 and R_2, eliminates imbalance due to bias current, but not due to unequal bias currents as were discussed in Section 8.9, Chapter 8.

11.2.1 Voltage Gain

The voltage gain of the amplifier is:

$$\frac{v_o}{v_s} = -\frac{A}{1 + \dfrac{R_1}{R_2}(1 + A)}$$

where v_o and v_s are respectively the output and input voltages, A is the open-loop operational amplifier gain, and R_1 and R_2 are the external input and feedback resistors, respectively.

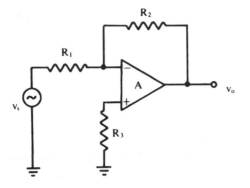

Approximate equations for practical design:

$$\text{Gain} = -\frac{R_2}{R_1}$$

Input resistance $= R_1$

Output resistance $= R_o \dfrac{R_2}{AR_1}$

Figure 11-1   The inverting amplifier.

The voltage gain equation is derived by neglecting the bias current and equating the current through R_1 to the current through R_2. If v_a is the potential with respect to ground at the inverting terminal, this gives $(v_s - v_a)/R_1 = (v_a - v_o)/R_2$. Again neglecting the bias current, the voltage at the noninverting input terminal is at ground. The output voltage v_o is then related to the voltage v_a through the open-loop gain A of the amplifier by $v_o = -Av_a$. Substituting v_a in terms of v_o from this last equation into the previous equation, we obtain the inverting amplifier gain equation given above.

Note that the closed-loop voltage gain (v_o/v_s) is a function of frequency since the open-loop gain A is a function of frequency. (See Figures 10-1, 10-2, and 10-3.) At low frequencies where $(R_1/R_2)(1 + A) \gg 1$ and $A \gg 1$ the closed-looped gain equation becomes for all practical purposes

$$\frac{v_o}{v_s} = -\frac{R_2}{R_1}$$

This is usually the frequency range in which the amplifier is used, and therefore this last equation is the common design equation. Practical constraints limit permissible values of R_1 and R_2, making an arbitrary large gain impossible. Required input impedance imposes a lower limit on R_1, and offset bias current limits upper values of R_1 and R_2. These constraints are taken into consideration in the design example shown in Section 12-2.

At high frequencies for which $(R_1/R_2)(1 + A) \ll 1$, the closed-loop voltage gain becomes $(v_o/v_s) = -A$. That is, at high frequencies the closed-loop voltage gain becomes equal to the open-loop voltage gain. This is illustrated in Figure 10-1. The magnitude of the gain changes from R_2/R_1 to A at a frequency f_B at which the corresponding "straight-line approximation" lines intersect. In reality, the transition from one gain to the other is gradual, and at f_B the gain is down approximately by 3 dB from the low-frequency gain. (If the open-loop gain curve is composed of two linear sections of 0 dB/dec and -20 dB/dec as exemplified by the response curve in Figure 8-3, then at the intersection of the "straight-line approximation" of the closed-loop and open-loop lines the gain is down *exactly* by 3 dB.)

11.2.2 Input Resistance

In practical designs, the input resistance of the amplifier is equal to the resistance of the externally connected resistor R_1.

$$R_{in} = R_1$$

For this to be true, the open-loop gain of the operational amplifier must be in the tens of thousands or larger; the open-loop output resistance of the operational amplifier must be about 100 Ω or less; and the open-loop input resistance of the amplifier (i.e., the resistance between the two input terminals) must be at least two orders of magnitude larger than R_1. All these conditions are commonplace with IC operational amplifiers.

11.2.3 Output Resistance

For the values of open-loop gain and input resistance of practical operational amplifiers, the output resistance of the inverting amplifier is

$$R_{out} = \frac{R_o \left(1 + \dfrac{R_2}{R_1} \right)}{A}$$

where R_o is the open-loop output resistance of the operational amplifier. The value of R_o is typically on the order of a hundred ohms. Typically, A is in the hundreds of thousands or millions, and R_2/R_1 is on the order of ten or one hundred. We see that R_{out} is a small fraction of R_o. The output resistance can be approximated in terms of the *loop gain* A_{Loop} $= A(R_1/R_2)$ as

$$R_{out} = \frac{R_o}{A_{Loop}}$$

For R_o = 200 Ω, A = 100,000, and R_2/R_1 = 10: this gives R_{out} = 0.02 Ω. However, caution must be exercised, since R_{out} can increase significantly if the loop gain for any reason (e.g., high frequency) goes down. For example, if R_o = 200 Ω, A = 10,000, and R_2/R_1 = 200, then R_{out} = 4 Ω.

11.2.4 Common-Mode Effect

The output of the amplifier including the effect of the common-mode input is

$$V_o = -\frac{R_2}{R_1} \left[v_s + \left(1 + \frac{R_1}{R_2} \right) \frac{v_{cm}}{CMRR} \right]$$

The expression $1 + R_1/R_2$ is close to unity, showing that the common signal appears attenuated by a factor of *CMRR* before it is amplified by the gain of the amplifier. The output voltage equation is correct for large open-loop gain A. At high frequencies for which A is small, the gain of the closed-loop amplifier approaches $-A$ (rather than

$-R_2/R_1$ given in the equation), and the relative effect of v_{cm} versus v_s increases, since *CMRR* decreases with decreasing A.

The common-mode rejection property reduces the effect of spurious or undesired signals reaching both input terminals of the operational amplifier. It does not help to reject undesired signals originating in the signal source or reaching the terminal to which the signal source is connected, as such signals do not possess the *common-mode* aspect, but appear as input signals.

11.3 THE NONINVERTING AMPLIFIER

The noninverting amplifier configuration is shown in Figure 11-2.

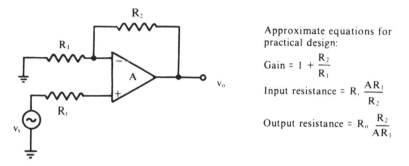

Approximate equations for practical design:

$$\text{Gain} = 1 + \frac{R_2}{R_1}$$

$$\text{Input resistance} = R_i \frac{AR_1}{R_2}$$

$$\text{Output resistance} = R_o \frac{R_2}{AR_1}$$

Figure 11-2   The noninverting amplifier.

11.3.1 Voltage Gain

The voltage gain of the amplifier is

$$\frac{v_o}{v_s} = \frac{A}{1 + \dfrac{R_1 A}{R_1 + R_2}}$$

where all the symbols are defined in the figure. The gain is a function of frequency since A is a function of frequency. (See Figures 8-3 and 10-1.)

The gain equation is derived by neglecting the bias current and equating the current through R_1 to the current through R_2. If v_a is the potential at the inverting input terminal with respect to ground, we have $-v_a/R_1 = (v_a - v_o)/R_2$. The output voltage is given by the open-loop gain A of the amplifier and the voltage between the inverting and noninverting terminals as $v_o = -A(v_a - v_s)$, where we neglected bias current and therefore any voltage drop across R_3. Solving for v_a using the last equation and substituting into the previous equation, the gain equation of the noninverting amplifier is obtained as stated above.

At low frequencies for which $R_1A/(R_1 + R_2) \gg 1$, the gain equation becomes

$$\frac{v_o}{v_s} = 1 + \frac{R_2}{R_1}$$

The gain in this range cannot be less than unity. The amplifier is ordinarily used in a frequency range where this equation is true, making the equation the standard design equation. Note, however, that at high frequencies where $R_1A/(R_1 + R_2) \ll 1$, the gain of the amplifier becomes A, equal to the open-loop gain of the operational amplifier. The transition from one gain to the other occurs at frequency f_B at which $A = 1 + R_2/R_1$ as shown in Figure 10-1. The discussion concerning the transition given for the inverting amplifier can be adapted to the noninverting amplifier.

11.3.2 Input Resistance

For the practical conditions of high gain A, high open-loop input resistance and low output resistance of the open-loop operational amplifier, the input resistance of the noninverting amplifier is

$$R_{in} = \frac{R_1A}{1 + \frac{R_2}{R_1}}$$

This can be written as

$$R_{in} = R_i A_{Loop}$$

where A_{Loop} is defined as the loop gain equal to $AR_1/(R_1 + R_2)$. In the frequency region where the amplifier is ordinarily used, the loop gain is very large and the input impedance R_{in} of the noninverting amplifier is several orders of magnitude larger than the open-loop input resistance R_i of the amplifier. Extraordinary high input resistances can be implemented. For example, if $R_i = 10^{12}$ Ω and $A = 200,000$ (typical for FET operational amplifiers) and the closed-loop gain of the amplifier is 100, then the loop gain of the amplifier is $200,000/100 = 2,000$ and the input resistance is 2×10^{15} Ω. At high frequencies the open-loop gain decreases causing the input impedance to decrease, as can be computed from the expression for R_{in} given above.

11.3.3 Output Resistance

For the practical conditions of very large A and R_i the output resistance is

$$R_{out} = \frac{R_o \left(1 + \frac{R_2}{R_1} \right)}{A}$$

This is the same expression as for the output resistance of the inverting amplifier. Everything that was said there is also true for the noninverting amplifier.

11.3.4 Common-Mode Effect

The output of the amplifier, including the effect of the common-mode input is

$$v_o = \left(1 + \frac{R_2}{R_1} \right) \left(v_s + \frac{v_{cm}}{CMRR} \right)$$

for high open-loop gain A. For low values of A, (e.g., at high frequencies) the gain of the noninverting amplifier is A, not $1 + R_2/R_1$ as given in the equation. The effect of v_{cm} becomes more pronounced as $CMRR$ decreases with decreasing A. As was also true for the inverting amplifier, the common-mode rejection property reduces the effect of spurious or undesirable signals reaching both input terminals of the operational amplifier, but not spurious or undesirable signals originating in the signal source or reaching the terminal to which the signal source is connected, as such signals do not possess the *common-mode* aspect, but appear as input signals.

11.4 CHOICE BETWEEN INVERTING AND NONINVERTING CONFIGURATIONS

If a gain less than unity is needed, the inverting amplifier must be used, since the gain of the noninverting amplifier is either equal to, or greater than, unity. If a very high input resistance is needed, the noninverting amplifier must be used. Its input resistance is equal to the product of the open-loop input resistance R_i of the operational amplifier, and the loop gain of the amplifier, both of which have high values. The input resistance of the inverting amplifier is equal to the resistance connected between the signal source and the inverting terminal. If this resistance is chosen high to have a high input resistance, and the feedback resistance for a gain greater than unity is even higher, the resistance values reach magnitudes too high for several reasons: they may introduce intolerable noise and cause unbalance due to unbalanced bias currents.

For multiple input signals, the inverting amplifier is normally more useful. For two input signals, as shown in Figure 11-3(a), the output is

$$v_o = -R_3 \left(\frac{v_{s1}}{R_1} + \frac{v_{s2}}{R_2} \right)$$

If $R_2 = R_1$ then

$$v_o = -\frac{R_3}{R_1} (v_{s1} + v_{s2})$$

Each input contributes its share to the output, independent of the other input circuit. The same concept holds true for three or more input signals.

The output of the noninverting amplifier (Figure 11-3(b)) is

$$v_o = \left(1 + \frac{R_2}{R_1} \right) \left(\frac{R_3 R_4}{R_3 + R_4} \right) \left(\frac{v_{s1}}{R_3} + \frac{v_{s2}}{R_4} \right)$$

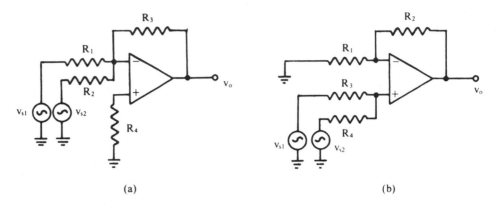

Figure 11-3 Multiple-input amplifiers. (a) Inverting amplifier. (b) Noninverting amplifier.

The contribution to the output of an input signal is determined not only by its own circuit, but also by the circuit associated with the other input signal. For example, the output associated with v_{s1} depends not only on R_1, R_2, and R_3, but also on R_4, which is associated with the circuit of v_{s2}. The resistor R_4 may represent in addition to an external resistor also the output resistance of the signal source v_{s2}. Thus, the factor by which a signal is amplified depends on other signal sources connected to the amplifier.

If $R_3 = R_4$, then

$$v_o = \left(1 + \frac{R_2}{R_1} \right) \frac{1}{2} (v_{s1} + v_{s2})$$

The gain of the amplifier is only one-half the gain for a single input. If there are three input signals, the gain is reduced by three, and if there are n input signal sources, the gain is reduced by a factor of n. This gain reduction does not exist in the inverting amplifier. Further, the high-input resistance aspect of the noninverting amplifier is lost, since the input resistance $R_{in}A_{Loop}$ is now paralleled with the resistors associated with the other input signal sources.

11.5 THE DIFFERENTIAL AMPLIFIER

The schematic diagram of a differential amplifier is shown in Figure 11-4(a). The amplifier accepts signals simultaneously at both the inverting and noninverting input terminals. A unique property of the differential amplifier is that it rejects common-mode signals originating at the signal source. It thus rejects 60-cycle-induced voltages picked up by electrodes pressed against the skin in biological or medical measurements; it rejects signals resulting from drifts in a previous amplification or transducer stage; and it rejects unwanted signals originating in so-called ground loops. The output signal is the amplified difference between the two input signals,

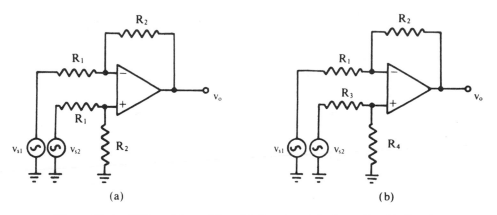

Figure 11-4   Differential amplifier. (a) Signals weighed equally. (b) Signals weighed unequally.

$$v_o = -\frac{R_2}{R_1}(v_{s1} - v_{s2}) = -\frac{R_2}{R_1}v_s$$

where v_s is the voltage applied between the two input terminals and we set $v_{s1} = v_s/2$ and $v_{s2} = -v_s/2$.

In Figure 11-4(a), equal sets of resistors are chosen for both input circuits. In general, the resistors can be different (Figure 11-4(b)), resulting in an output of the weighted input signals.

$$v_o = \frac{R_4(R_1 + R_2)}{R_1(R_3 + R_4)}v_{s2} - \frac{R_2(R_3 + R_4)}{R_1(R_3 + R_4)}v_{s1}$$

This equation reduces to the previous one for the special case:

$$R_3 = R_1 \text{ and } R_4 = R_2$$

11.5.1 Common-Mode Effect

The differential amplifier rejects common-mode signals appearing at the operational amplifier input terminals as well as those originating or appearing at the signal source. The output in the presence of a common-mode voltage at the input terminals of the operational amplifier is

$$v_o = -\left[\frac{R_2}{R_1}(v_s) + \left(1 + \frac{R_2}{R_1}\right)\frac{v_{cm}}{CMRR}\right]$$

For a common-mode voltage at the source the output voltage is

$$v_o = -\frac{R_2}{R_1}\left(v_s + \frac{v_{cm}}{CMRR}\right)$$

The relative contribution of the common signal v_{cm} to the output is small because it is divided by *CMRR*. As the frequency increases, the *CMRR* deteriorates (decreases) and the relative contribution of v_{cm} increases.

11.6 THREE-RESISTOR FEEDBACK NETWORK

An inverting amplifier with a three-resistor feedback network is shown in Figure 11-5(a).

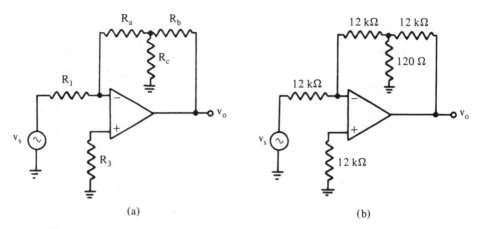

Figure 11-5   Inverting amplifier with three-resistor feedback circuit. (a) General schematic diagram. (b) Example.

The equivalent feedback resistance R_2 of this amplifier is

$$R_2 + \frac{R_a R_b + R_a R_c + R_b R_c}{R_c}$$

The voltage gain of the amplifier in terms of the equivalent resistance is

$$\frac{v_o}{v_s} = -\frac{R_2}{R_1}$$

The resistor R_3, as usual, is used to balance bias current effects and is equal to the parallel combination of R_1 and R_2. The advantage of using such an arrangement is illustrated by the following example.

Example

Determine the voltage gain of the amplifier shown in Figure 11-5(b). The effective feedback resistance is

$$R_2 = \frac{12 \times 10^3 \times 12 \times 10^3 + 12 \times 10^3 \times 120 + 12 \times 10^3 \times 120}{120}$$

$$= 1.224 \times 10^6 \ \Omega$$

The voltage gain is

$$\frac{v_o}{v_s} = -\frac{1.224 \times 10^6}{12 \times 10^3} = -102$$

Had we used the amplifier configuration of Figure 11-1 with $R_1 = 12\ \text{k}\Omega$ (to maintain the same input impedance), then to have a gain of 102, the feedback resistor R_2 in Figure 11-1 would have to be $12\ \text{k}\Omega \times 102 = 1.224\ \text{M}\Omega$. The advantage of the circuit of Figure 11-5 is evident in that the highest resistor value of $12\ \text{k}\Omega$ is used in the circuit as compared to $1.224\ \text{M}\Omega$ to obtain the same gain and the same input impedance. The use of relatively low resistance values reduces the noise associated with resistors and also suppresses the effect of stray wiring capacitances.

11.7 AC-COUPLED AMPLIFIERS

At times it is necessary to remove the DC component from the previous stage or signal source before the signal is processed further. This is done by AC coupling.

11.7.1 Inverting Amplifier

The circuit is shown in Figure 11-6(a). The configuration is the same as that of the DC inverting amplifier, except that a capacitor is inserted to block the DC voltage. The size of the capacitor is determined by the lowest frequency that must be amplified. The gain of an inverting amplifier is in general terms of impedance Z_1 and Z_2 (Figure 11-6(b)),

$$\frac{V_o}{V_s} = -\frac{Z_2}{Z_1}$$

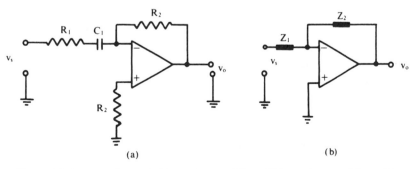

(a)   (b)

Figure 11-6   (a) AC-coupled inverting amplifier. (b) Inverting amplifier with impedance networks.

In the circuit of Figure 11-6(a), $Z_2 = R_2$ and $Z_1 = R_1 + 1/j2\pi fC_1$. The transfer function is therefore

$$\frac{V_o}{V_s}(j2\pi f) = -\frac{j2\pi fR_2C_1}{1 + j2\pi fR_1C_1}$$

It is seen that for $f = 0$, $(V_o/V_s) = 0$; for $f \to \infty$, $(V_o/V_s) \to -R_2/R_1$; the 3-dB break frequency is $f_{3dB} = \frac{1}{2}\pi R_1C_1$. For high frequencies ($f > f_{3dB}$) the capacitor is essentially a short circuit and the input and output impedances are identical to those of the DC-coupled inverting amplifier. The input impedance, for example, is therefore equal to R_1.

    There will be a DC offset output voltage caused by the combination of the DC input offset voltage and the input offset current. Because of the presence of C_1, the DC voltage between the input terminals will appear at the output with a gain of unity. Thus, the offset voltage at the output is

$$v_{o(offset)} = v_{i(offset)} + R_2I_{(offset)}$$

    Note that the resistor connected to the noninverting terminal to minimize the offset caused by the bias currents is equal to the feedback resistor R_2; not to the parallel combination $R_2\|R_1$ as would be the case for a DC-coupled amplifier, since the capacitor blocks the bias current, which is therefore supplied to the inverting terminal only through the feedback resistor.

11.7.2 Noninverting Amplifier

An AC noninverting amplifier is shown in Figure 11-7. The gain of the amplifier as function of frequency is

$$\frac{V_o}{V_s}(j2\pi f) = \frac{1}{1 + \dfrac{1}{j2\pi fR_3C_1}} \times \frac{1 + j2\pi f(R_2 + R_3)C_2}{1 + j2\pi fR_2C_2}$$

The capacitor C_2 can be replaced with a short circuit, in which case the compensation resistor R_1 at the noninverting terminal is replaced with the parallel combination $R_1 = R_2\|R_3$ to minimize bias effects. The gain equation is then

$$\frac{V_o}{V_s}(j2\pi f) = \frac{1}{1 + 1/j2\pi fR_1C_1}\left(1 + \frac{R_3}{R_2}\right)$$

For high frequencies ($f > 1/2\pi R_1C_1$) the transfer function reduces to that of the DC-coupled noninverting amplifier,

$$\frac{V_o}{V_s} = \left(1 + \frac{R_3}{R_2}\right)$$

    Note that it is necessary to provide DC paths to both input terminals for bias currents. It is therefore not possible to remove R_1 from the circuit, which would make the circuit analogous to the DC-coupled noninverting amplifier with the associated high-input

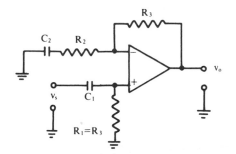

Figure 11-7   AC-coupled noninverting amplifier.

impedance. With R_1 present, the input impedance is limited to R_1. A high-input impedance, noninverting amplifier is presented in the next section.

11.7.3 High-Input-Impedance AC-Coupled Noninverting Amplifier

A "bootstrapped" high-input-impedance AC-coupled amplifier is shown in Figure 11-8.
Bias current to the inverting terminal is supplied through R_3, and to the noninverting
terminal through R_1 in series with R_2. At low frequencies C_2 provides high impedance. It
is interesting to see what happens at higher frequencies ($f > \frac{1}{2}\pi(R_1 + R_2)C_2$), where C_2
provides essentially a short circuit. With C_1 of the same order of magnitude as C_2 or
larger, at higher frequencies the voltage drop across C_1 approaches zero, causing R_2 to be
effectively connected between the input terminals of the amplifier. The voltage across R_2
is then only the very small offset voltage, which implies a very small current through R_2,
and therefore a high-input resistance.

$$R_i = v_s/i_s$$

As usual, this is true, provided the open-loop gain of the amplifier is very high, causing v_2
$\approx v_4$. At very high frequencies, for which the open-loop gain of the amplifier becomes

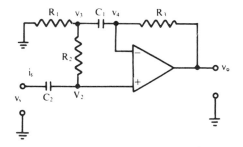

Figure 11-8   High-input-impedance AC-coupled amplifier.

small, the voltage across the amplifier input terminals increases, and substantial currents can flow through R_2, causing a reduced input impedance.

For high frequencies, for which the capacitors offer essentially no reactance,

$$v_o/v_s = 1 + \frac{R_3}{R_1}$$

The input impedance is

$$Z_{in} = \frac{1}{j2\pi fC_2} + R_1 + R_2 + j2\pi fC_1R_1R_2$$

At low frequencies, the input impedance is high because of C_2 and at high frequencies because of C_1 except, as noted earlier, the input impedance decreases again at frequencies for which the open-loop gain of the amplifier decreases.

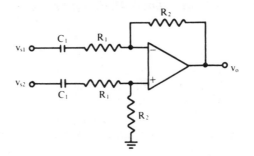

Figure 11-9   AC-coupled differential amplifier.

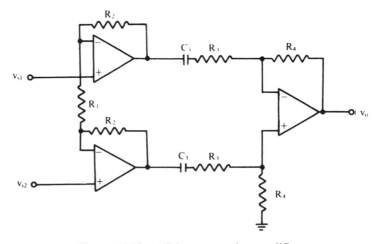

Figure 11-10   AC instrumentation amplifier.

11.7.4 Differential Instrumentation Amplifiers

Figures 11-9 and 11-10 show, respectively, an AC differential amplifier and an AC high-input resistance instrumentation amplifier. The output voltage of the differential amplifier, Figure 11-9 is

$$v_o = \frac{j2\pi f R_2 C_1}{1 + j2\pi f R_1 C_1} (v_{s2} - v_{s1})$$

At high frequencies the equation reduces to $v_o = (R_2/R_1)(v_{s2} - v_{s1})$

The output of the instrumentation amplifier (Figure 11-10) is given by

$$v_o = \left(1 + \frac{2R_2}{R_1}\right) \left(\frac{j2\pi f R_4 C_3}{1 + j2\pi f R_3 C_3}\right)(V_{s2} - V_{s1})$$

The DC decoupling is done between the first and the second stages so that the signal source is not loaded by a capacitor. At high frequencies, the output equation reduces to

$$v_o = \frac{R_4}{R_3} \left(1 + \frac{2R_2}{R_1}\right)(v_{s2} - v_{s1})$$

12

Design Examples and How to Avoid Pitfalls

12.1 INTRODUCTION

In this chapter we illustrate how information presented in previous chapters is utilized in the design of real-life systems, where specifications must be met within given tolerances. Those design examples show you how to apply knowledge and data to meet particular needs, and they serve as guides for meeting other design objectives.

12.2 VOLTAGE OPERATIONAL AMPLIFIER

It is required to design an amplifier to change the voltage level of a signal of maximum rms value of 300 mV supplied by a source with 5-Ω output impedance. The maximum load current that the signal source can supply is 500 μA. The output of the amplifier must have a maximum rms value of about 6 V feeding into a 5-kΩ load. The maximum frequency of the signal is 18 kHz. The amplifier must operate from 0 to 50°C with a distortion not to exceed 1% of maximum output.

The statement that the amplifier output must be "about" 6 V means that the actual gain need not have a precise, specified value (perhaps an adjustable gain exists elsewhere in the system), which implies that the resistors that determine the gain need not be precision resistors; 5% resistors are adequate. We shall try to design the amplifier around the basic configuration of Figure 11-1. The resistance ratio R_2/R_1 that determines the gain of the amplifier must be 6 V/300 mV = 20. In decibels the gain is 20 log 20 = 26 dB. We consider the popular amplifier LM 741 (see data sheets in Appendix A2).* An examina-

*Identical to LM 747 characteristics.

tion of the open-loop transfer characteristic shows for a closed-loop gain of 20, a 3-dB frequency of about 50 kHz, and a peak-to-peak output swing at 18 kHz of 16 V, while our need is for $2 \times \sqrt{2} \times 6 = 16.97$-V peak-to-peak output. Furthermore, a check of the required slew rate shows that we need $SR \geq 2\pi f V_o = 2\pi \, 18 \times 10^3 \times 6 \times 10^{-6} = 0.68$ V/μs. The typical slew rate of the LM 741 is 0.5/μs. Thus, the use of this operational amplifier is ruled out. In terms of operational amplifiers the frequency requirement is severe, and an uncompensated amplifier may be the right choice. Examining the LM 748 (see data sheets in the Appendix), shows external compensation effected by means of a capacitor connected between two terminals. For a 3-pF capacitor the 3-dB frequency for a 26-dB closed-loop gain is about 500 kHz, and the output peak-to-peak voltage at 18 kHz is 29 V. The slew rate of the amplifier is 0.8 V/μs. These values are for a power supply of ± 15 V and a temperature of 25°C, as read off the curves. The minimum gain over the range -55 to 125°C is 93 dB. A shift of the gain frequency-response curve from the typical 100-dB (DC) gain to 93 dB shows that the amplifier still meets our requirements, and there is ample margin, since the required temperature range is limited to 0 to 50°C. The output current swing for a peak-to-peak output voltage of 17 V is about 52 mA peak-to-peak at 25°C and 32 mA peak-to-peak at 125°C. Our need is for only $2 \times \sqrt{2} \times (6/5000) = 0.0034$ A or 3.4 mA peak-to-peak. We use, therefore, the LM 748.

One percent of 6 V is 60 mV. Therefore, to ensure that the distortion not exceed 1% of maximum output, unwanted voltages at the input must be limited to 60 mV/20 = 3 mV. The *maximum* input offset voltage over the range of -55 to 125°C for the LM 748, and 0 to 70°C for the LM 748 C is specified as 6.0 mV and the average coefficient of input offset voltage is 3.0 μV/°C. Thus, if we balance the amplifier to zero output at 25°C, then at the extremes of 0°C and 50°C the offset voltage will be 3 μV/°C $\times$ 25°C = 75 μV, which is considerably below the 3-mV limit. This provides a high degree of confidence that the design will be valid although the 3.0-μV/°C figure given in the data sheet is an *average* value. (A curve is not provided by the particular manufacturer.)

The minimum value of the resistor R_1 is imposed by the signal source specification of 300 mV and 500 μA: $R_{1 \, \text{min}} = 300 \times 10^{-3}$ V/500×10^{-6} A = 600 Ω. The maximum value is determined by the input offset current. The error introduced by the bias current is compensated for by adding a resistor R_3 between the noninverting terminal and ground. However, the two bias currents are not exactly the same, and for the temperature range 0 to 70°C the difference in current, i.e., the offset current, is specified as at most 300 nA. We can express this in terms of a voltage by noting that the bias current flows through R_1 and R_2 in parallel:

$$\frac{v_a}{R_1} + \frac{v_a}{R_2} = i_{\text{offset}}$$

$$i_{\text{offset}} \frac{R_1 R_2}{R_1 + R_2} = v_a$$

but $R_2 = 20 R_1$, giving

$$i_{\text{offset}} \times 0.95 R_1 = v_a$$

As a guide, we want to ensure that $v_a \leq 0.3$ mV, which is 10% of maximum total unwanted voltage permissible to ensure not more than 1% distortion of maximum output. Thus, for the upper limit,

$$300 \times 10^{-9} \times 0.95\, R_1 = 0.3 \times 10^{-3}$$

from which

$$R_1 \leq 1053 \ \Omega$$

Thus, we choose $R_1 = 1{,}000 \ \Omega$ and $R_2 = 20$ kΩ. These are standard 5% resistors in the required range giving a gain of 20. To balance the bias current, we connect a resistor $R_3 = 910 \ \Omega$ (a standard 5% resistor approximately equal to the parallel combination of R_1 and R_2) to the noninverting terminal. The voltages at the input terminals resulting from the bias currents are given by $I_{bias} \times R_3 = 0.3$ mV. This is small in comparison with the specified* common-mode voltage swing of ± 12 V.

The minimum common-mode rejection ratio of the operational amplifier is specified as 70 dB or 3,162. From the common-mode effect equation in Section 11.3.4, we find that a common-mode signal of 0.3 mV at the input terminals of the operational amplifier contributes to the output.

$$\left(1 + \frac{R_2}{R_1} \right) \frac{v_{cm}}{CMRR} = \frac{21}{3162}\, 0.3 \times 10^{-3} = 2 \times 10^{-6} \ \text{V}$$

This is $(2 \times 10^{-6}/6) \times 100 = 3 \times 10^{-5}$% of maximum output.

The resistance values of the resistors used in the circuit are small enough for resistor noise not to pose a problem.

The complete circuit of the amplifier is shown in Figure 12-1. The voltage offset balancing circuit, consisting of P, R_4, and R_5 (all carbon-type components), is taken from the data sheet. The 3-pF ceramic capacitor is used for open-loop gain-frequency compensation (or phase compensation) for stability as given in the data sheet. The 0.1-μF capacitors are tantalum electrolytic decoupling capacitors to filter any disturbances from other parts of the circuit that could be transmitted through the power supplies.

12.3 TRANSCONDUCTANCE OPERATIONAL AMPLIFIER

It is required to design an amplifier with a closed-loop voltage gain A_{CL} of 6, and an input resistance R_S of 25 kΩ to drive a load resistance R_L of 10 kΩ. The maximum input signal voltage is 100 mV peak-to-peak. The power supply voltages are ± 6 V, and the temperature is 25°C. The circuit that will be used for the design is shown in Figure 9-3(a). Our main task, which differs from the design procedure that uses a voltage operational amplifier, is the design of the bias circuit to provide the required open-loop transfer characteristic. We shall demonstrate the design by a step-by-step procedure.

We shall use the CA3060 amplifier, the characteristics of which as supplied by the

*Obtained from a manufacturer's data sheet.

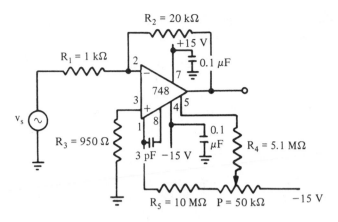

Figure 12-1 Design example.

manufacturer, are shown in Appendix A4. We shall make use of the bias regulator supplied inside the IC package to minimize the voltage and current offset, depending on supply voltage variations. The use of the bias regulator causes more current drain from the power supply and must be avoided if current drain is critical.

The input offset voltage V_{IO}, the bias current I_B, and the input offset current I_{IO} all increase with an increase in the amplifier bias current I_{ABC}. Thus, to keep the total effective offset voltage, $V_{equ.\ offset} = V_{IO} + R_S I_{IO}$, as small as possible, I_{ABC} should be kept as small as possible, large enough only to provide the needed transconductance g_m and sufficient output swing.

Calculation of required g_m. The open-loop voltage gain A_{OL} of the amplifier should be at least 10 times larger than the closed-loop voltage gain A_{CL} to allow for sufficient feedback.

$$A_{OL} = 10\ A_{CL} = 10 \times 6 = 60$$

But

$$g_m = \frac{A_{OL}}{R_L} = \frac{60}{10\ \text{k}\Omega} = 6\ \text{mS}$$

Calculation of I_{ABC}. From the curve supplied by the manufacturer, we find that for $g_m = 6$ mS, we need $I_{ABC} = 30\ \mu$A (using the "minimum" curve for g_m). From the I_{OM} curve, we find that the corresponding minimum output current is 60 μA. We shall check below to see if this is sufficient current for our design.

Input resistor R_S. The value of R_S is determined by the specification of the required input resistance of the amplifier.

$$R_S = 25\ \text{k}\Omega$$

Calculation of feedback resistor R_F.   The closed-loop gain of an inverting amplifier is given as $A_{CL} = R_F/R_S$. Hence,

$$R_F = A_{CL} R_S = 6 \times 25 \text{ k}\Omega = 150 \text{ k}\Omega$$

Calculation of R_B.   The bias resistor R_B is used to minimize voltage offset resulting from bias currents just as in the voltage-operational amplifier.

$$R_B = R_S \| R_F = \frac{R_S R_F}{R_S + R_F} = \frac{25 \text{ k}\Omega \times 150 \text{ k}\Omega}{25 \text{ k}\Omega + 150 \text{ k}\Omega} = 21.4 \text{ k}\Omega \simeq 22 \text{ k}\Omega$$

Calculation of total output current.   The output current is the sum of the current through the load R_L and the feedback resistor R_F. The output voltage v_o is equal to the voltage gain times the signal input voltage v_s.

$$v_o = 6 \times 0.1 \text{ V} = 0.6 \text{ V peak-to-peak}$$

The total current is therefore

$$i_o = i_L + i_F = \frac{v_o}{R_L} + \frac{v_o}{R_F} = \frac{0.6 \text{ V}}{10 \text{ k}\Omega} + \frac{0.6 \text{ V}}{150 \text{ k}\Omega} = 64 \text{ }\mu\text{A peak-to-peak}$$

This corresponds to a peak current of $64/2 = 32$ μA, which is well below the 60-μA figure obtained from the data sheet for $I_{ABC} = 30$ μA.

Calculation of R_Z and R_{ABC}.   The bias regulator contains two transistors and a Zener diode. To ensure that the bias regulator provides good regulation, the Zener diode current should be equal to at least $1.5 I_{ABC} = 1.5 \times 30$ μA $= 45$ μA. This ensures that the Zener diode remains in its voltage saturation (control) region. The Zener diode can be considered to be connected in series with R_Z and the power supplies. The Zener voltage is 0.7 V. The value of R_Z is then obtained as follows:

$$1.5 I_{ABC} \times R_Z = V^+ + V^- - 0.7 \text{ V}$$

$$R_Z = \frac{6 \text{ V} + (-6 \text{ V}) - 0.7 \text{ V}}{45 \text{ }\mu\text{A}} = 111 \text{ k}\Omega \simeq 110 \text{ k}\Omega$$

The value of R_{ABC} is determined by the required I_{ABC}:

$$R_{ABC} = \frac{V_1 - V_6}{I_{ABC}}$$

where the subscripts 1 and 6 stand for the terminal numbers in Figure 9-2(a).

The values of V_1 (bias regulator voltage) and V_6 (amplifier bias voltage) are obtained from the bias regulator characteristics shown in Figure 9-2(b). We see that for $I_{ABC} = 30$ μA, the curves give $V_1 = 6.8$ V and $V_6 = 0.625$ V giving

$$R_{ABC} = \frac{6.8 - 0.625}{30 \text{ }\mu\text{A}} = 206 \text{ k}\Omega \simeq 200 \text{ k}\Omega$$

Summary of design.   The circuit is as shown in Figure 9-3(a) with the following values of components:

$$R_S = 25 \text{ k}\Omega$$
$$R_F = 150 \text{ k}\Omega$$
$$R_B = 22 \text{ k}\Omega$$
$$R_L = 10 \text{ k}\Omega$$
$$R_Z = 110 \text{ k}\Omega$$
$$R_{ABC} = 200 \text{ k}\Omega$$

The 0.1-μF capacitors are used to decouple the circuit from the power supplies.

12.4 PITFALLS TO AVOID

Essentially, to avoid poor designs, the designer must be very familiar with the operational amplifier specifications, and take into consideration those characteristics that may affect critically the design specifications. For example, in the first design example in Section 12.2 we saw that certain limits were imposed on the input resistor R_1 by the input offset current. The objective was to ensure that the cumulative percentage distortion error not exceed the specified percentage. The designer must be flexible. Often, it is possible to allow a larger contribution to the error by one parameter, while decreasing the contribution by another parameter to obtain an overall compatible system. If voltage regulation or common-mode signals contribute potential problems, their effects must be considered, as was explained in Sections 8.7 and 8.6.1, Chapter 8. In low-level amplifier designs, noise becomes an important consideration. Noise curves as given in Appendix A, p. 157 must be consulted.

In noninverting amplifiers, unlike inverting amplifiers, the voltage levels of the inverting and noninverting input terminals of the operational amplifier are not kept near ground, but at the signal level. An operational amplifier must be chosen whose common-mode range is equal to, or exceeds, the signal magnitude.

Special precautions must be taken with the transconductance amplifier. Both the gain-frequency response and the slew rate are very sensitive to stray as well as load capacitance. This is a result of the high impedances in the circuit. For example, a 10-kΩ load resistance and an associated stray capacitance of only 15 pF produces a break frequency of 1 MHz as computed from $f = \frac{1}{2}\pi RC$. This is shown in Figure 9-2(b).

The slew rate $SR = dv_{out}/dt$ is given by the maximum possible output current I_{om} and the capacitance C_L that must be charged: $SR = I_{om}/C_L$. Since the currents in a transconductance amplifier are in the microamp range, even relatively small values of capacitances will cause poor slew rates. For example, if $I_{om} = 10$ μA and $C_L = 100$ pF, the slew rate is only 0.1 V/μs.

Part III PASSIVE COMPONENTS

13
Resistors

13.1 INTRODUCTION

Resistors can be divided into two main classes: (1) conventional bulk components, and (2) thick- and thin-film resistor arrays. Bulk components provide the largest variety, are stocked in large quantities by local suppliers of electronic components, and are the most convenient to use in developmental designs and small-quantity production. There are four major types of bulk resistors: carbon-composition resistors, wirewound resistors, metal-film resistors, and carbon-film resistors. Each type has unique advantages and certain limitations. Thick- and thin-film resistor arrays have been made available for general use in some variety since the early 1970s. They have the advantage of being contained in standard DIPs (dual-in-line packages) and SIPs (single-in-line packages), making their assembly into a circuit easy and giving the completed circuit the appearance of an assembly of packages rather than individual components. This makes for simple assembly and should be considered for larger production quantities.

13.2 CARBON-COMPOSITION RESISTORS

Carbon-composition resistors are very reliable and are the least expensive of all resistor types. Consequently, they are the most widely used in electronic circuits. They are available from 1 Ω to 22 MΩ with tolerances of 3%, 5%, 10%, and 20%, and power ratings of ⅛, ¼, ½, 1, and 2 W. The temperature coefficient (*TC*) of carbon-composition resistors is relatively high and can be considered to be an average of 0.1%/°C (or 1000 ppm/°C) over the applicability range up to a temperature of about 160°C. However, in the normal range of application between 0 and 60°C the coefficient is considerably less. It increases rapidly below 0°C and above 60°C. The noise level of carbon-composition resistors is the highest among all resistors.

13.3 WIREWOUND RESISTORS

Wirewound resistors are divided into three broad categories: high-power, high-accuracy, and general-purpose resistors. Since these resistors are made of wire coils, their inductance is relatively high and must be taken into account at higher frequencies. *Bifilar*-wound resistors, however, are available (at a higher cost) to minimize the inductive effect. They are usually referred to as *noninductive wirewound resistors.*

 High-power wirewound resistors are generally available within the 1-Ω-to-100-kΩ resistance range with 5-to-200-W dissipation, and tolerances of 5% and 10%. Although integrated circuits generally operate at very low power levels, power resistors must be used, for example, in the output stage of an audio amplifier in conjunction with power transistors which are, in turn, fed from an integrated low-power amplifier stage.

 High-accuracy wirewound resistors exhibit a low temperature coefficient and excellent long-term stability. Most commonly used are 1% resistors, but resistors with *tolerances* as low as 5 ppm are available. Resistors with *temperature coefficients* as low as 5 ppm/°C are also available. Most commonly used power ratings are ¼, ½, and 1 W, but 3-, 5-, 10-, 25-, and 50-W resistors are also available, and resistance values are from 1 Ω to 100 kΩ. High-accuracy resistors are used in applications such as precision-active full-wave rectifiers or other circuits where precision voltage divider networks are needed.

 General-purpose wirewound resistors are available in the ¼-Ω-to-10-kΩ resistance range and power ratings of ½, 1, and 3 W. They duplicate, to some degree, application areas of carbon-composition resistors and are used mainly when low temperature coefficients are needed.

13.4 METAL-FILM RESISTORS

Metal-film resistors are suitable for the most demanding applications. Their resistance ranges from 0.1 Ω to 1.5 MΩ. Resistors with temperature coefficients as low as 1 ppm/°C are available with excellent long-term stability and power ratings of ¹⁄₁₀, ⅛, ¼, ½, and 1 W. These resistors have extremely low noise figures, which make them desirable for low-level preamplifiers and other low-noise applications.

13.5 CARBON-FILM RESISTORS

Two unique characteristics of *carbon-film* resistors are the high available resistance values (10 Ω to 100 MΩ) and the negative temperature coefficient of resistance. The resistance tolerances are 0.5% and higher. They have low noise in resistance values below 100 kΩ and are inexpensive. These resistors are used where high values are needed, as well as for temperature compensation in some circuits where advantage is taken of the negative temperature coefficient.

13.6 THIN- AND THICK-FILM RESISTOR ARRAYS

The era of integrated circuits gave rise to the use of packaged resistor arrays. These are assemblies of individual flat-shaped DIP packages with 14 (or other specified number) leads for external connections. By interconnecting external leads, any number of resistors can be connected in series or in parallel to obtain a variety of values. For information on available values the reader should contact the manufacturers. Many are listed in the *Electronic Engineers Master (EEM)* publication.

The main advantages of these resistor packages are their compactness, neatness, and ease of assembly into an electronic system. They should be considered particularly in large quantity runs, where cost savings may be realized.

13.7 THERMISTORS

A thermistor is a resistor with a high negative temperature coefficient of resistance (*TCR*). This property can be used for temperature compensation in electronic networks. (The device can also be used as a sensor for temperature measurements.)

Thermistors are fabricated by sintering certain combinations of metallic-oxide semiconductors. The *TCR* can be as large as $-5\%/°C$ over a specified temperature range.

Example

A precision instrument requires a reference resistor of about 1 MΩ. While the exact value of resistance is not critical, once it is fixed, it must not change. A 1-MΩ metal-film resistor is incorporated for this purpose. The design is such that component self-heating is negligible, and the environment temperature can cause a fluctuation of 5°C inside the instrument. The *TCR* of the resistor is 1 ppm/°C or 0.0001%/°C. Over a span of 5°C the resistance of the 1-MΩ resistor will change by

$$0.0001\%/°C \times 5°C = 0.0005\%$$

or by

$$1{,}000{,}000 \ \Omega \times 0.000005 = 5 \ \Omega$$

To compensate for this change, a 40-Ω thermistor is connected in series with the resistor. In the given temperature range, the *TCR* of the thermistor is

$(-)2.5\%/°C$. The change in resistance of the thermistor over the 5°C temperature range is therefore,

$$-40 \ \Omega \times 0.025/°C \times 5°C = -5 \ \Omega$$

compensating for the resistance change of the 1-MΩ resistor.

13.8 PITFALLS TO AVOID

Nominal resistor values have been chosen and standardized by manufacturers to avoid waste in manufacturing. That is, any resistor manufactured subject to process control limitations falls within the tolerances of a nominal value. Accordingly, resistors with ±20% tolerance are available with the following nominal values: 10, 15, 22, 33, 47, 68, 100 Ω and in multiples of 10 of these values; resistors with ±10% tolerance are available with the nominal values of 10, 12, 15, 18, 22, 27, 33, 39, 47, 56, 68, 82, 100 Ω and multiples of 10 of these values; and ±5% tolerance resistors are available with the nominal values of 10, 11, 12, 13, 15, 16, 18, 20, 22, 24, 27, 30, 33, 36, 39, 43, 47, 51, 62, 68, 75, 82, 91, 100 Ω and multiples of 10 of these values. Tighter tolerance resistors are often specified in fractions of ohms and manufacturers' or suppliers' catalogs should be consulted.

In designing circuits, the designer should specify *nominal* resistor values to avoid delay or confusion or poor choice of resistors by the technician assembling the circuit.

Power dissipation of a resistor of value R is given by I^2R or V^2/R, where I is the current through the resistor, and V is the voltage across the resistor. It must be noted that resistor power ratings are given for stated ambient temperatures. If no value for the ambient temperature is stated, it is normally safe to assume that the power rating is valued up to about 70°C.

The power rating is limited by the maximum allowable temperature of the resistor. Too high a temperature can destroy a resistor, but even more troublesome to the maintenance person is a permanent change in the resistance value. Too high a temperature can cause an irreversible change of resistance resulting from a change in the resistor material composition. This can cause, for example, a permanent change in amplifier gain or oscillator frequency.

To avoid such problems it is necessary to *derate* the power rating of a resistor when it is to be used at higher temperatures. If no curves or other data for derating are available, it is usually reasonable to derate the power rating of the resistor linearly from 100% to 0% rating between 70 and 150°C according to the formula:

$$\text{Power rating at } t°C = 100\% \text{ power rating } \left(1 - \frac{t - 70}{80}\right)$$

Thus, for example, the power dissipated in a ½-W resistor should not exceed 375 mW if the circuit is to be operated at an ambient temperature of 90°C.

It must also be borne in mind that resistors are *noisier* at higher temperatures. The thermal noise in a resistor results from the random motion of electrons in the resistor, and increases with temperature. The thermal noise manifests itself in a voltage across, and a corresponding current through, the resistor. This voltage increases in proportion to the square root of the absolute temperature, and is in proportion to the square root of the resistance. The absolute temperature is obtained by adding 273° to the temperature in degrees Celsius. The noise voltage contains all frequencies. It is thus possible to reduce the circuit noise level by filtering the output, allowing only frequencies in the signal frequency range to reach the output.

The temperature coefficient of resistance must be considered at high ambient temperatures or at high power levels when high component temperature may be reached. This is, however, of no major concern in most designs using ICs, except for output stages, since the power levels are normally very low, considerably below the resistor power ratings. It must also be noted that the change in system performance as a result of a change in resistance values is less than the change in resistance if the performance depends on resistance *ratios*. If, for example, the gain of an amplifier is given by the ratio $R_2/R_1 =$ 100 K/5 K $= 20$, the *TC* is 1%/°C and the temperature rise is 20°C, then the gain at the higher temperature is 120 K/6 K $= 20$, so that no change in gain has taken place. On the other hand, if the 5-K resistor is connected to the input terminal of an operational amplifier, then the voltage drop across the resistor caused by the bias current will change as a result of the resistance change. Unless this voltage drop is very small in comparison with the voltage drop caused by the signal, a change in amplifier performance will take place.

If the performance of a circuit depends on an *RC* time constant, the performance will, in general, change as a function of temperature, since both *R* and *C* are temperature-dependent. If the circuit must operate over a wide temperature range this can be of concern. It is then necessary to choose components with small *TC*s. It is interesting to note that high degrees of temperature stability can be obtained by proper choices of resistors and capacitors with *TC*s of nearly equal magnitudes but opposite signs. For example, a carbon-film resistor in combination with a low-loss ceramic capacitor can be used. Both components have *TC*s of the order of 100 ppm, the coefficient being negative for the resistor and positive for the capacitor. Some degree of compensation for resistive networks can be obtained by connecting a carbon-composition resistor with a positive *TC* in series with a carbon-film resistor with a negative *TC*.

A word of caution about *high-voltage limitations* is necessary. The voltage levels usually encountered in IC design are of the order of 5, 15, or 30 V, and the voltage, except as it relates to power dissipation, is no cause for concern. However, an IC may be used, for example, in the feedback circuit of a high-voltage power supply to stabilize the voltage, and the designer may have to assume the responsibility for the complete system. The high voltage of hundreds or thousands of volts may cause an irreversible change or the destruction of a resistor resulting from the high electric field before the maximum power dissipation is reached. It is, therefore, necessary to consider the voltage rating of a resistor in addition to the power rating. A typical value for wirewound resistors is 1,000 V. When unusually high voltages are encountered, the manufacturer's catalog must be consulted.

High-frequency effects on resistors must be considered in applications above 200 kHz. Care must be exercised since the effective value of a resistor can be off by several times the nominal DC value. Wirewound resistors, except the more-expensive bifilar resistors, must be avoided because of the inductive effect that adds impedance to the total impedance of the circuit at higher frequencies. But even carbon-composition resistors deviate from their low-frequency value at higher frequencies because of the shunt stray capacitance that lowers the impedance. Typical frequency characteristics of carbon-composition resistors are shown in Figure 13-1.

High-resistance valued resistors above 100 MΩ can be a source of trouble if not handled correctly. The leakage shunt resistance between the terminals lowers the effective resistance, particularly at high relative humidity or in the presence of contaminants on the resistor surface. High-valued resistors must be cleaned in a solvent before they are connected into a circuit.

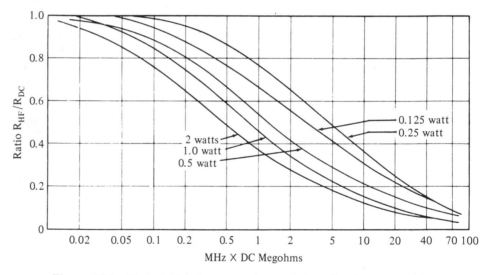

Figure 13-1   Typical high-frequency characteristics of carbon-composition resistors.

13.9 HOW TO CHOOSE THE RIGHT TYPE RESISTOR

The general rule is: choose the least-expensive resistor that will perform the required function satisfactorily. The price of the resistor will usually increase with tighter tolerance and *TC* specifications. In a large variety of applications, carbon-composition resistors will meet the requirements. For highest precision, wirewound resistors must be used, and in high frequencies where the inductance is objectional, the bifilar-wirewound resistors must be used. Very high values can be obtained only in carbon-film resistors, and very low values are available in metal-film resistors.

13.10 POTENTIOMETERS AND RHEOSTATS (VARIABLE RESISTORS)

A *potentiometer* is a three-terminal device with a constant resistance between two terminals, but the resistance between the third terminal and either of the two other terminals is variable. It is used as a voltage divider. A *rheostat* is a device with a variable resistance between two terminals. It is possible to use a potentiometer as a rheostat by connecting the variable terminal to either of the fixed terminals. This is common practice in electronics (see Figure 13-2). In this book we use the term *potentiometer* for either device.

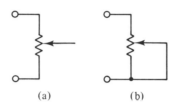

(a)   (b)

Figure 13-2   Potentiometer and rheostat. (a) Potentiometer. (b) Rheostat.

13.10.1 Potentiometers (Variable Resistors)

Miniature Potentiometers

The miniaturization of electronic circuits stimulated the development of a large variety of *miniature* and so-called *subminiature* potentiometers. They are convenient to use and have become the standard component, except for front-panel mounting.

Potentiometers can be divided into three types, according to the material of the resistive element: carbon, cermet, and wirewound potentiometers. In addition to resistance values and power ratings, specified parameters include resistance tolerance and minimum resistance (which are of no importance except when the potentiometer is used at one of the extreme positions, which would indicate a poor choice of potentiometer, and is not the normal case), linearity (which is also of no great importance, since the miniature potentiometers are set once for the proper value and are not controlled by the equipment user), resolution, and temperature coefficient of resistance.

The carbon-type potentiometer is least expensive, but it has the widest tolerances and the largest *TC*. Next in price comes the cermet type, and finally, the wirewound type. The latter is the most expensive type, but it has the tightest specifications. Miniature potentiometers come in rectangular and circular shapes and a variety of ratings. Terminals are in the form of insulated stranded leads, solder lugs, or printed-circuit pins. Especially convenient for circuits using ICs are dual-in-line potentiometers that have the standard DIP size (TO-116) compatible with automatic insertion equipment. Electronics suppliers' catalogs list many types, and further details can be obtained from manufacturers' data sheets.

Carbon types are available from 5 Ω to 1 MΩ, cermet types from 10 Ω to 100 kΩ.

Power ratings range from 0.2 to 1 W. Temperature coefficients can be as low as 100 ppm/°C, resistance tolerance as low as ±3%, and linearity ±0.15%. High resolutions are achievable with multiturn potentiometers. Pots with 25, 20, 15, 10, and 4 turns are available; they cost several times more than single-turn pots, which can be used for less demanding circuits. They all have screwdriver adjustments. Normally, these miniature pots are mounted directly on the assembly board, although panel-mount units with heavier screw-adjustable shafts and mounting nuts are available for most types at a higher cost.

Examples of typical potentiometer specifications are:

1. wirewound 0.5 W at 70°C, 25-turn adjustment, absolute minimum resistance 0.7% or 0.5 Ω, whichever is greater, for 10 Ω to 50 kΩ, 0-to-5% for 100 kΩ. Maximum operating temperature 125°C, resistance tolerance 10%, sealed against sand, dust, and humidity. Size: $5/16 \times 1/4 \times 1/4$ in.;
2. cermet 0.75 W at 25°C; 0 W at 125°C, infinite resolution; stability ±0.05%; resistance tolerance ±10%, operating temperature range −55 to +125°C; size: $0.350 \times 0.19 \times 0.75$ in.

Panel-Mount Potentiometers

Most of the microminiature potentiometers described in the previous section are also available for panel mounting, but for ruggedness and for convenience for operator's control, larger-size potentiometers are usually used for panel mounting. Most common are $3/4$-in.-diameter 2-W pots rated at 70°C, with $1/4$-in.-diameter shaft for either knob or screwdriver control.

Panel potentiometers are available with carbon-composition, cermet, and wirewound elements. For general purpose, single-turn (270-deg. rotation) pots are available, and for precision control, 5-turn or 10-turn pots are used. Readily available carbon-element potentiometers range from 100 Ω to 1 MΩ, cermet elements range from 50 Ω to 5 MΩ, and wirewound elements range from 10 Ω to 300 kΩ. The power ratings range from $1/2$ to 50 W.

Examples are:

1. hot-molded carbon element, $1/2$-in. diameter, $1/2$ W at 70°C, operating temperature −55 to +120°C, resistance tolerance ±10%, $1/8$-in.-diameter shaft;
2. cermet element, $1/2$-in. diameter, 1.0 W at 125°C, resistance tolerance ±10%, $1/8$-in.-diameter shaft;
3. precision 10-turn pot, 1-in. diameter, 2.0 W at 70°C, maximum operating temperature 125°C, resistance tolerance ±3%, linearity ±0.2%, $1/4$-in.-diameter shaft. Count dials, digital readout, and clock-face readout dials for easy control-setting are also available.

Taper refers to the relation between the fraction of the resistance to the shaft position. In a linear-taper pot the resistance is proportional to the shaft position. Other tapers include a logarithmic taper that is used commonly in the audio output stages to conform to the logarithmic sensitivity of the ear to sound amplitude. Sinusoidal and other special tapers are also available in right-handed or left-handed configurations.

13.10.2 Pitfalls to Avoid

Note that the *power rating* derates with temperature. The rating is specified for a given temperature derating to zero power at a given higher temperature. If no graphical data are given, a linear derating curve between the two temperatures can be assumed.

Most important is to note that the power dissipation is given by I^2R, where I is the current and R is the resistance. The rating is specified for the *total resistance* of the potentiometer. If only a fraction of the resistance is used in the circuit, the permissible power dissipation must be reduced to the same fraction of the specified power rating.

The term *infinite resolution* can be misleading when applied to practice. Carbon-composition and cermet element pots have infinite resolution, since in contrast with wirewound elements, the elements are continuous in the direction of wiper movement. However, if the potentiometer is a one-turn pot, the resistance depends very critically on the exact position of the wiper, and settings to an exact value are very difficult to obtain. On the other hand, in multiturn pots the resistance changes relatively little with shaft position, and needed resistance value can be obtained relatively easily, even if the element is wirewound and has finite resolution.

A high degree of *linearity* is important only if the resistance value (or corresponding gain, etc.) is set by a dial reading. If, on the other hand, the value is set by a voltage reading, the extra cost for better linearity is not justified.

14

Capacitors

14.1 INTRODUCTION

Selecting the "right" capacitor from among the great variety of available types can be a confusing task for the circuit designer. Capacitors are characterized by voltage rating, insulation resistance, dissipation factor, frequency response, size, and, of course, capacitance. In addition, the temperature dependence of several of these parameters can be quite large. For a particular application we must weigh the importance of each factor and select the capacitor that most closely meets our needs. A casual selection may result in degraded circuit performance, and overspecification will certainly result in increased cost and size.

A capacitor with one or two high-quality specifications can cost more than many ICs and will probably be larger. With the exception of the ICs, few circuit components require as much attention to detail in their specifications as does the capacitor.

14.2 CAPACITOR SPECIFICATIONS

Each capacitor type (ceramic, film, electrolytic, etc.) has sufficiently different characteristics and areas of applicability to warrant individual discussion. But first, let us examine and clarify some of the specifications applied to the capacitor.

14.2.1 Capacitance

The specification seems almost trivial, but an understanding of what gives a capacitor its capacitance will aid in understanding many other characteristics. All capacitors consist basically of two metal electrodes separated by an insulator which is called the dielectric (see Figure 14-1). In the ceramic capacitor, for example, this structure is realized by depositing a metal film on both sides of a fired clay disk and attaching leads. A metalized polycarbonate capacitor consists of two plastic films with metal deposited on one side of each, laid on top of each other, and wound in a spiral to form a tubular shape. In any case, Figure 14-1 is an appropriate mechanical representation. The capacitance C of this structure is given by

$$C = \frac{\epsilon_o \, KA}{d} \tag{14-1}$$

where ϵ_o = permittivity of free space, A = area of either plate, K = dielectric constant of the dielectric material, and d = distance between the plates.

To increase the capacitance for a given size, we could reduce d and, by using more layers, increase A. But this process is limited by the maximum field strength rating of the dielectric, since for a given voltage, the electrical field increases as we decrease the thickness.

We have no control over ϵ_o, but K, the dielectric constant, can range from one to several thousand for various insulating materials and is, therefore, an important parameter. Almost all capacitance variations with temperature, frequency, and applied voltage are due to variations in K.

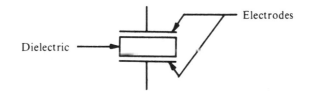

Figure 14-1   Schematic diagram of capacitor.

14.2.2 Insulation Resistance (DC Leakage)

This specification is a measure of the capacitor's ability to block DC currents. It is defined by measuring the current I_L flowing through the capacitor when a DC voltage V_{DC} is applied across the terminals and calculating the equivalent resistance $R_S = V_{DC}/I_L$. The effect of this resistance can be modeled by the circuit of Figure 14-2. Looking at Equation 14-1, with ϵ_o, K, and d fixed, the capacitance increases with increased surface area A and, since the leakage flows in the insulating material, the leakage current increases proportionally and the equivalent shunt resistance R_S decreases. For a given type capacitor, C, and given voltage rating, we find that the product $R_S \times C$ is constant. This number is called the "megohm-microfarad product," and since its units are seconds, it is the time constant of the capacitor C and its shunt resistance R_S. Typical high-quality film capacitors have time constants, $R_S \times C$, in excess of 50,000 s (14 h). At low capacitance values, this relationship implies extremely high R_S values, and in practice, manufacturers specify $R_S \times C$ and a maximum value of R_S, which need not be exceeded.

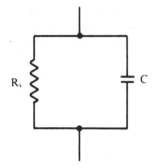

Figure 14-2   Equivalent circuit of capacitor considering insulation resistance.

For long-term integrators and sample hold applications, the insulation resistance is the important specification.

Example

Find the equivalent parallel resistance R_S of a 1-μF polycarbonate capacitor whose megohm-microfarad product is 50,000 s.
Solution:

$$C \times R_S = 50,000 \text{ s, so}$$

$$R_S = \frac{50,000 \text{ s}}{1 \text{ μF}} = 50,000 \text{ MΩ}$$

Electrolytic capacitors have a voltage-sensitive insulation resistance, and normally manufacturers specify the actual DC leakage current at the rated voltage instead of the equivalent resistance.

14.2.3 Dissipation Factor, DF, Quality Factor, Q, and Equivalent Series Resistance, ESR

The dissipation factor is a measure of the power losses in a capacitor under AC conditions. In a few capacitors, this loss is partially composed of $I^2 \times R$ losses in the electrodes, but for most capacitors, all the AC losses occur in the dielectric material. This effect results from the microscopic motion and resultant heating of small elements of the dielectric caused by the varying forces associated with the alternating electric field. The dissipation constant is defined as the total energy lost in the capacitor, divided by the total energy stored in the capacitor, where the exciting waveform is a single cycle of a sine wave. Dissipation factor is normally expressed as a percentage of stored energy. For a general purpose ceramic capacitor, the *DF* is approximately three percent. For high-quality capacitors, typical values are on the order of tenths of a percent. It is important to note that the *DF* is the loss per cycle, and even though the *DF* is relatively constant over a broad frequency range for many materials, the total power loss increases linearly with frequency. Thus, for a capacitor C used in a sine wave application at frequency f, we have:

$$\text{Power loss} = \tfrac{1}{2}\, Cv^2 \times DF \times f$$

where v = rms values of the sine wave.

The Q factor is the inverse of the dissipation factor: $Q = 1/DF$. If $DF = 0.01$, $Q = 100$, and if $DF = 0.1\%$ then $Q = 10{,}000$. The DF and Q factors are of interest when designing high Q resonant tanks and other AC applications.

We can model these losses by placing an equivalent resistance in series with the capacitor as in Figure 14-3. However, since DF is relatively constant, the equivalent series resistance, ESR, is frequency-dependent. Specifically:

$$ESR = \frac{DF}{2\pi f C}$$

where f = the operating frequency.

Since the power loss in electrolytic capacitors is a combination of electrode resistance and dielectric losses, most manufacturers publish a graph of ESR as a function of frequency, as the DF is not constant in these types.

$$ESR = \frac{DF}{2\pi f C}$$

Figure 14-3   Equivalent circuit of capacitor considering the dissipation factor.

14.3 TYPES OF CAPACITORS

With the above background, let us now discuss some major families of capacitors commonly used in low voltage level electronic circuits.

14.3.1 Ceramic Capacitors

Surely, the ceramic capacitor family has the widest range of applicability of all capacitor types. Suppliers have divided this family into three groups based on the dielectric constant K of the ceramic material.

Low K—Temperature compensating and NPO capacitors. These capacitors are fabricated with low-K material and are the high-performance members of this family. Unfortunately, the low K results in large size for a given voltage rating and capacitance, so these types are limited to about 5000 pF. These types are available in a zero-temperature coefficient of capacitance—NPO—and several negative temperature coeffi-

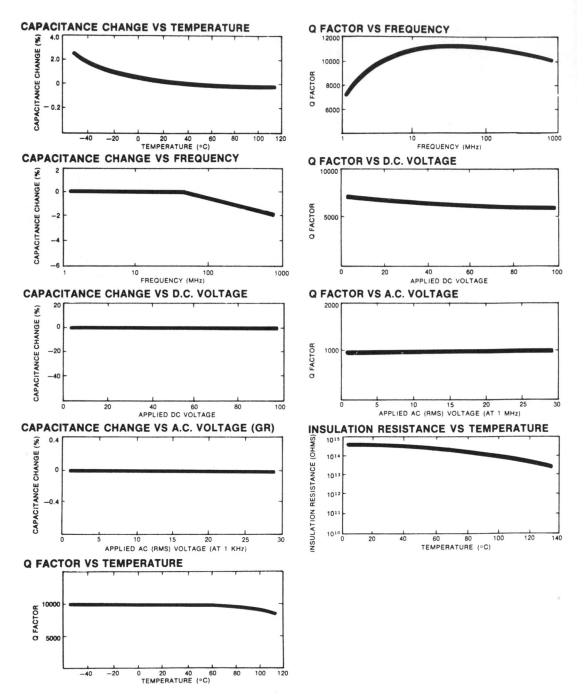

Figure 14-4 Typical performance characteristics of low-*K* NPO capacitors. (Reprinted with permission of Murata Corp. of America.)

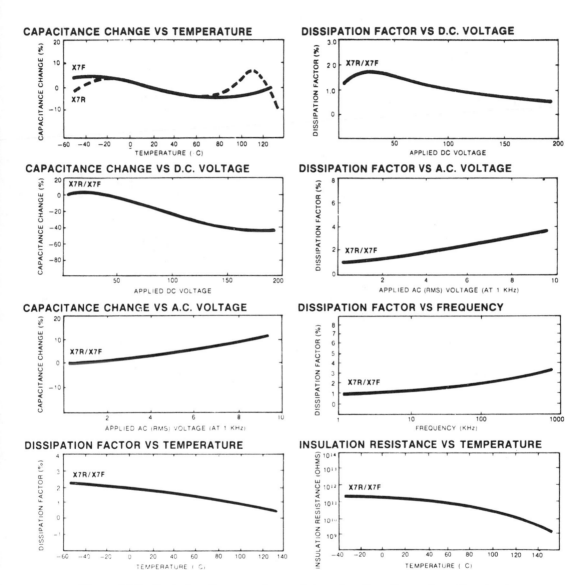

Figure 14-5 Typical performance characteristics of medium-*K* capacitors. (Reprinted with permission of Murata Corp. of America.)

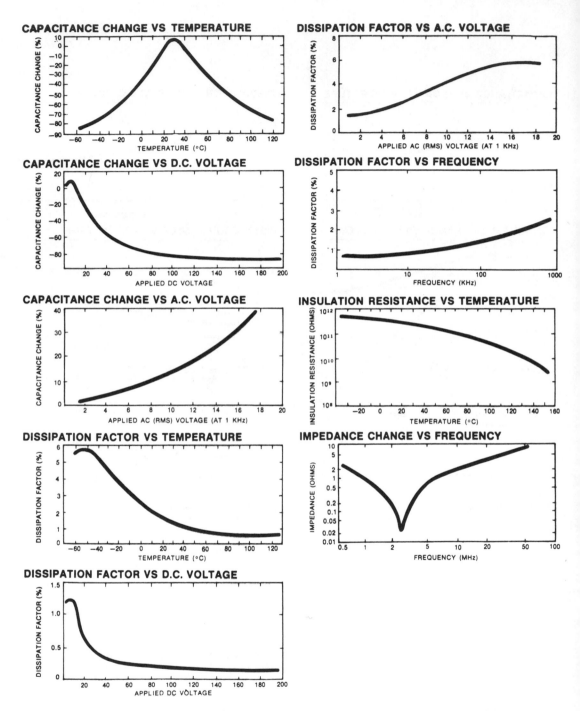

Figure 14-6 Typical performance characteristics of high-*K* capacitors. (Reprinted with permission of Murata Corp. of America.)

cients—N220, N750, etc. The curves of Figure 14-4 demonstrate the excellent performance of this group. Most notable is the stability with parameter variation and high Q factors. These capacitors are excellent choices in RF and precision applications at low capacitance values.

Medium K—Temperature stable. This is an intermediate formulation and is frequently advertised as temperature stable, although looking at Figure 14-5 we see that the performance is significantly worse than the low-K types in all areas. Of course, the higher K value results in more capacitance for a given size. Typical designations of medium-K capacitors are X7F, X7R, and Y5R.

High K—General purpose. This group is advertised as "general purpose for noncritical applications." Looking at Figure 14-6 we see that the performance of this group can only be described as poor. The advantage, of course, is very high capacitance in a small package, but if we are working near the voltage rating or at a temperature extreme, we may not have as much capacitance as we had hoped for.

14.3.2 Film-Type Capacitors

The dielectric in this family is a plastic film. The electrodes can be sheets of aluminum or evaporated aluminum deposited directly on the film in a high vacuum (metalized). Typically, tapes of film and metal are alternated and wound in a tubular shape. Depending on the electrode-terminating techniques, this construction can exhibit rather high or very low series inductance. Table 14-1 provides a very good performance comparison among the various film types.

Polyester (Mylar*). Except for working voltage, this type has somewhat poorer specifications at 25°C than other members of this group. Notable, however, is its wide useful temperature range. Typical performance characteristics are shown in Figure 14-7.

Polystyrene. For low-voltage applications and operating temperatures below 100°C, polystyrene is an excellent dielectric. Unfortunately, the low dielectric constant and low working voltage imply that for the same capacitance and voltage rating, the polystyrene capacitor is much larger than other film capacitors. This type has a predictable $(-)$ 120 ppm/°C temperature coefficient. Extremely high Q factors at 25°C over a broad frequency range recommend this material for high-Q resonant tanks and filters.

Polycarbonate. This type offers good performance over a wide temperature range.

Polypropylene. at the expense of capacitance stability with temperature, polypropylene offers moderate temperature range performance similar to polystyrene in a somewhat smaller package.

Teflon. Teflon is a very-high-insulation-resistance material with otherwise moderate performance to 200°C.

14.3.3 Mica Capacitors

Mica capacitors outperform the low-K ceramics at very high ($>$200 MHz) frequencies. Unfortunately, the mica's small dielectric constant results in capacitors with low capaci-

---

*DuPont registered trademark.

TABLE 14-1 Dielectric Performance Comparisons for Film Capacitors

	TEMPERATURE °C								
	−55	−30	0	25	45	65	85	105	125
DIELECTRIC CONSTANT @ 1 khZ (Typical)									
Mylar*	—	—	—	3.1	—	—	—	—	—
Polystyrene	—	—	—	2.5	—	—	—	—	—
Polycarbonate	—	—	—	2.8	—	—	—	—	—
Polypropylene	—	—	—	2.4	—	—	—	—	—
Aluminum Oxide	—	—	—	7.0	—	—	—	—	—
CAPACITANCE CHANGE % @ 1 kHz (Typical)									
Mylar*	−5.0	−2.0	−1.0	0	1.0	1.5	2.0	7.00	15.00
Polystyrene	1.0	0.7	0.3	0	−0.3	−0.6	−0.9	—	—
Polycarbonate	−1.0	−0.30	0	0	0.0	0.0	0	−0.13	0.20
Polypropylene	+1.4	+1.2	+0.6	0	−0.4	−1.6	−2.5	−3.50	—
IR MEGOHM-MICROFARADS (WVDC 2 min. Typical)									
Mylar*	$>10^5$	$>10^5$	$>10^5$	10^5	4×10^4	7×10^3	10^3	4×10^2	7×10
Polystyrene	$>10^6$	$>10^6$	$>10^6$	10^6	7×10^5	5×10^5	4×10^4	0	0
Polycarbonate	$>10^6$	$>10^6$	$>10^6$	4×10^5	3×10^5	2×10^5	10^5	5×10^4	10^3
Polypropylene	$>10^5$	$>10^5$	$>10^6$	10^5	4×10^4	7×10^3	8×10^2	10^2	—
DISSIPATION FACTOR % @ 1 kHz (Typical)									
Mylar* @ 1 µF	1.0	1.10	0.70	0.32	0.16	0.10	0.15	0.52	0.96
Polystyrene @ 2000 pF	0.07	0.08	0.04	0.03	0.05	0.07	0.08	∞	∞
Polycarbonate @ 0.1 µF	0.45	0.30	0.20	0.07	0.08	0.08	0.10	0.15	0.16
Polypropylene @ 2000 pF	0.06	0.06	0.06	0.06	0.06	0.06	0.07	0.06	—
DIELECTRIC ABSORPTION % (Typical)									
Mylar*	—	—	—	0.20	—	—	—	—	—
Polystyrene	—	—	—	0.02	—	—	—	—	—
Polycarbonate	—	—	—	0.08	—	—	—	—	—
DC WORKING VOLTAGE VOLTS/GA# (Typical)									
Mylar*	8	8	8	8	8	8	8	6	4
Polystyrene	3	3	3	3	3	3	3	0	0
Polycarbonate	5	5	5	5	5	5	5	5	5
Polypropylene	8	8	8	8	8	8	8	6	—

	FREQUENCY Hz @ 25°C				
	60	120	10^3	10^5	10^6
DISSIPATION FACTOR % @ 25°C					
Mylar* @ 0.1 µF	0.20	0.23	0.32	1.10	10.00
Polystyrene @ 500 pF	0.01	0.01	0.01	0.01	0.04
Polycarbonate @ 0.1 µF	0.05	0.06	0.07	0.16	4.00
Polypropylene @ 500 pF	0.01	0.01	0.01	0.01	0.13
Q FACTOR @ 25°C					
Mylar* @ 0.1 µF	500	435	313	91	10
Polystyrene @ 500 pF	10,000	10,000	10,000	10,000	2,500
Polycarbonate @ 0.1 µF	2,000	1,670	1,430	625	25
Polypropylene @ 500 pF	10,000	10,000	10,000	10,000	770

*DuPont registered trademark

114

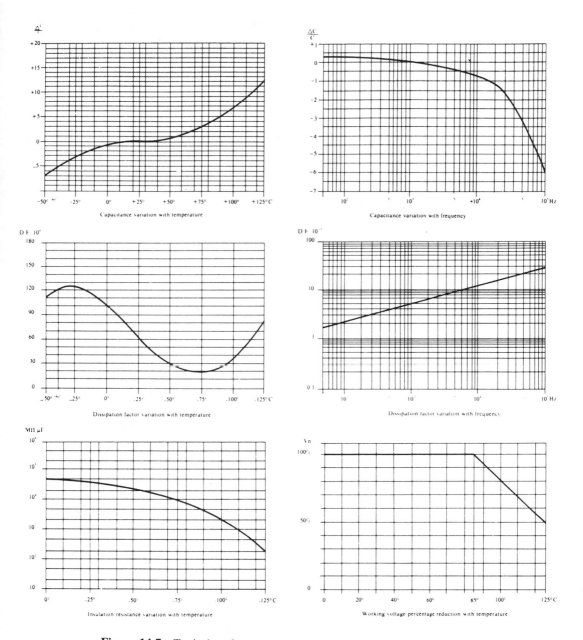

Figure 14-7 Typical performance characteristics of metalized polyester capacitors.

tance-to-volume ratios. The temperature coefficient can be controlled very closely by a proper choice of the type of mica and the method of construction. It is possible to manufacture mica capacitors with predictable temperature coefficients in the range from -200 ppm/°C to $+200$ ppm/°C. Thus, they can be designed to compensate for temperature dependencies of other components in the circuit, and, for example, oscillators can be built with minimum frequency drift with temperature. The dissipation factor is very small, on the order of 0.02% to 0.1%, and remains low into the high megahertz range.

Two kinds of mica capacitors are available: the *metal-foil* kind, which consists of alternate layers of mica and metal foil, and the *silvered-mica* kind, which consists of mica sheets with silver deposited directly on the mica. The latter type results in lower temperature coefficients and also in greater stability during operation.

14.3.4 Electrolytic Capacitors

The outstanding feature of the electrolytic capacitor is its high capacitance-to-volume ratio. Common electrolytic capacitors are available in the range from 1 µF to 0.1 F with voltage ratings up to 500 V. The higher capacitance values are available only at low-voltage ratings. For high-voltage, high-capacitance needs, several units must be paralleled. Shortcomings of electrolytic capacitors are their wide tolerance range from -20 to $+100\%$, their high equivalent series resistance (*ESR*), and their high DC leakage currents.

Electrolytic capacitors are listed as aluminum or tantalum types, dependent upon the electrode material. Both come as polar or nonpolar capacitors. The polar types are by far the more common. The voltage can be applied only with the specified polarity. If the polarity is reversed, the leakage current is excessive, causing overheating and possibly the explosion of the capacitor. The nonpolar electrolytic capacitor can be connected to a voltage with either polarity. The nonpolar capacitor, however, has only half the capacitance of a polar capacitor of the same volume. The dielectric in an electrolytic capacitor is an oxide grown on the electrode metal by anodization. In the polar capacitor only one electrode is anodized, while both electrodes are anodized in the nonpolar capacitor. If the wrong polarity has been connected momentarily to a polar capacitor, or if the capacitor has been stored unused for several months, it may be necessary to form the dielectric layer by the application of the DC working voltage with the proper polarity until the leakage current reduces to the nominal value.

Relatively small electrolytic capacitors are made in tubular form with one wire extruding from either end with the positive terminal clearly marked. Larger-valued capacitors can be contained in one can with or without one common electrode.

A miniature type of electrolytic capacitor is the solid tantalum type molded into an epoxy. Capacitance values range from a fraction of a microfarad to several hundred microfarads.

14.3.5 Trimmer Capacitors

Trimmer capacitors are variable capacitors used for a one-time screwdriver adjustment to an exact desired value. The adjustment is normally done experimentally to peak a resonant circuit, for example, or to adjust a band-pass filter, or a time constant. Commonly used are small screw-mounted ceramic capacitors on the order of ½ × ¾ in. in size. Capacitance values range from a picofarad to several hundred picofarads. The adjustable range of any one unit covers from 4 : 1 to 10 : 1. Examples are 1 to 4 pF, 8 to 50 pF, and 10 to 100 pF. Voltages are in the range from a few hundred to a few thousand volts with dissipation factors on the order of 0.2% at 1 MHz.

14.3.6 Feed-Through Capacitors

Feed-through capacitors are used in high-frequency circuits where it is desired to bypass to ground a wire passing through a shield box. The capacitor provides terminals on both sides of the wall and a capacitance to the wall. Two types are available: a soldered type and a thread-and-nut mounted type. Capacitance values range from 100 to 5000 pF.

14.3.7 Air-Variable Capacitors

These are used where a tunable resonance circuit is required, such as for the front-end of a receiver. It consists of interleaved sets of metal plates, one set stationary and the other set movable by means of a rotating shaft. The dielectric is air. By rotating the shaft, the capacitance can be varied between a few picofarads and a few hundred picofarads.

14.4 PITFALLS TO AVOID

Often, a high-valued capacitor is used to bypass relatively low frequencies. Such a capacitor, however, may not bypass high frequencies (which one would expect to be bypassed, based on circuit *RC* time constant consideration for an ideal capacitor), because the *ESR* of electrolytic capacitors becomes excessive at high frequencies. The remedy in such a case is to parallel the capacitor with a small-value (0.1-μF) high-frequency capacitor.

15

Inductors

15.1 INTRODUCTION

When designing an electronic circuit or system with ICs, one tries to avoid the use of inductors, because of their bulkiness. Tuned circuits can be designed using certain RC bridge networks, but at times inductors cannot be avoided. At very high frequencies, inductors are not a cause of much concern, because the few microhenries needed can be obtained by bending a wire into a few-turn, self-supporting air inductor. Important factors to consider are the inductance, the resistance, the resulting Q value, and the maximum DC current rating.

The DC current rating is meaningful only for iron-core inductors and corresponds to the current value that will drive the iron core into saturation, resulting in reduced inductance. The quality factor Q is the ratio of the coil reactance X_L, which depends on frequency, to the DC resistance R_{DC}:

$$Q = \frac{X_L}{R_{DC}} = \frac{\omega L}{R_{DC}}$$

where ω is the angular frequency and L is the inductance.

The effective AC resistance can also be a function of frequency as a result of eddy current and hysteresis losses in the core. This results in the broadening of the bandwidth of a tuned circuit and is sometimes used deliberately for this purpose.

15.2 AIR-CORE INDUCTORS

At very high frequencies, in the range of 100 kHz to 1 GHz, the needed inductances are usually on the order of a few tenths of a microhenry to several hundred microhenries. Inductors with these values can be obtained as *air coils*. Such coils are either self-supported or wound on insulating cylindrical forms. At the higher frequencies where very low inductance values are needed, the inductors can be handmade by forming a few loops of wire. We shall illustrate this with an example.

The inductance of an air coil is given by:

$$L = \frac{r^2 n^2}{9r + 10l}$$

where L = inductance, in microhenries;
r = radius of the coil, in inches;
l = length of the coil, in inches; and
n = number of turns of the coil.

Suppose we want to construct a coil of 2-μH inductance. The wire diameter is determined by the current that it must carry and by the required strength and rigidity (if it is self-supporting). Suppose that a wire diameter of 0.005 in. is suitable. The length of the coil can then be expressed as $l = 0.005n$. We choose a coil radius of 0.25 in. The equation for the inductance then becomes:

$$2 = \frac{0.25^2 \times n^2}{9 \times 0.25 + 10 \times 0.005 \times n} = \frac{0.125 \, n^2}{2.25 + 0.05 \, n}$$

Solving for n we obtain only one physically realizable solution, $n = 6.8$. (The other solution for n is negative.) Thus, a 7-turn coil is the solution. The coil can be stretched slightly lengthwise to obtain just the right value. In practice, the needed value would be obtained by expanding or compressing the coil slightly to obtain the correct functioning of the circuit in which the coil is being used, rather than by adjusting the coil independently.

15.3 IRON-CORE INDUCTORS

Iron-core inductors are available commercially in a variety of physical sizes and electric characteristics. Most desirable for designs with integrated circuits are small-size (or "miniature") inductors. Typical case sizes are less than ½-in. diameter and ¼-in. height. Inductance values at 0 VDC range typically from 0.1 to 120 mH, and DC resistance values range from a few ohms for the lower-valued inductors to several thousands of ohms for the higher-inductance coils. The maximum permissible DC current may be as low as 0.2 mA or as high as 90 mA. Values of Q of about 40 at 10 kHz are available. There are also available high-Q inductors at low frequency, but they are larger in physical size. For example, a 2-H inductor with a case size of 1.3 in. in diameter and 1.6 in. in height may have a Q value of 40 at 60 Hz.

16
Transformers

16.1 INTRODUCTION

Transformers are used in power supplies primarily to obtain the needed voltage levels, in IF and RF stages primarily for circuit tuning, in video and audio stages primarily for impedance matching, and in pulse circuits for voltage-pulse transmission. In addition to providing these primary functions, transformers can also provide electric isolation, polarity inversion, and impedance-level change for noise reduction and for other functions. It should be noted, however, that in modern design with ICs, the use of transformers is rather rare.

16.2 POWER-SUPPLY TRANSFORMERS

When deciding on the use of a power supply for a particular electronic circuit, the designer can choose between a ready-made packaged power supply with IC voltage regulator and the design of a power supply using individual components. The first choice results in the most expensive power supply, but it is ready for use with input terminals for connection to the 115-V line and with output terminals for the desired regulated DC voltage. A selection of regulated voltage levels and polarities, as well as current levels from a few milliamperes to several amperes, is available. Off-the-shelf IC voltage regulators are available up to a power output of 5 W. The packages must be mounted on a heatsink, and the designer must provide the unregulated DC voltage to the voltage regulator. This involves the design of a circuit that contains a power transformer, rectifiers, and capacitors. Finally, a similar circuit must be designed if the designer assumes the responsibility for the design of the complete regulated power supply.

The selection of a power transformer is based on voltage ratings and ratios, frequency, power rating, and winding resistance. The primary side voltage must correspond to the line voltage, which usually is 115 V, and the specified frequency must correspond to the line frequency, which usually is 60 Hz. A frequency of 400 Hz is common in some military applications. Transformers are available with a frequency-range specification from 50 to 5000 Hz, but usually the frequency specification is 50 to 60 Hz or 400 Hz. The voltage that can be supported by a transformer winding is given by the relation:

$$V = 4.44 f N \phi \times 10^{-8}$$

where V = the rms voltage (in volts) of a sinusoidal voltage
      f = the frequency (in hertz)
      N = the number of turns
      ϕ = the magnetic flux (in maxwells)

For square-wave applied voltages, the number 4.44 must be replaced with the number 4 and V is the peak voltage. The designer selecting a transformer need not be concerned with this equation, but it is convenient to use it to explain some simple, but important, facts. If the rated voltage is applied, but the frequency is too low, the flux ϕ will increase above the nominal value causing saturation of the magnetic core and overheating of the transformer. Also, the impedances of the coils will change with frequency.

The equation also shows that for a given value of AC flux, at a given frequency, the voltage across any given coil is proportional to the number of turns; the ratio between the voltages of any two coils is therefore proportional to the turn ratio. The secondary voltage of a power transformer must be chosen according to the desired value of the DC-regulated voltage. The peak value of the secondary voltage must be higher than the desired DC-regulated voltage. In some power transformers there are two secondary windings that can be connected either in series or in parallel, providing two possible values for the secondary voltage. Standard voltages are 5, 10, 12, 14, 20, 24, 27, 30, 33, 36, and 40 V rms. Thus, transformers are available for any desired DC-regulated voltage level suitable for ICs and transistors. Higher voltages are also available for transistors requiring hundreds of volts. These are usually listed under power transformers for vacuum tubes. The low-level voltage transformers are small, having linear dimensions on the order of 1 to 1.5 in., and often have lug connections suitable for printed circuit mounting.

The power rating is given either as a *VA* (voltage × current) rating or in terms of output current, which when multiplied by the output voltage gives the power. The value of the winding resistance is important, since it affects the regulation. To obtain the voltage output under full load, the product of the resistance times the full-load secondary current must be subtracted from the nominal output voltage. The resistance must be taken as the sum of the secondary winding resistance plus the product of the secondary to primary turn ratio squared times the primary resistance.

Since power transformers are used in conjunction with rectifiers, some manufacturers specify the output voltage in terms of the DC voltage when the transformer feeds either into a half-wave rectifier or into a full-wave rectifier. Thus, for example, the secondary voltage of a transformer may be rated at 9 VDC for a half-wave rectifier, and 18 VDC for a full-wave or bridge rectifier. These specifications apply to the circuits shown in Figures 16-1(a) and 16-1(b), respectively. Higher voltages may be obtained through the use of capacitor input filters. In this case, however, the rated DC current must be reduced by a factor of 2. If a voltage doubler is used, the current must be reduced by a factor of 4.

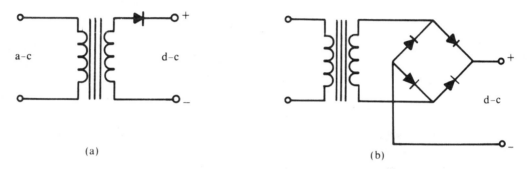

(a)   (b)

Figure 16-1   Voltage rectifier circuits for power supplies. (a) Half-wave rectifier. (b) Full-wave rectifier.

16.3 AUDIO TRANSFORMERS

The important characteristics of audio transformers are the power rating, frequency response, and impedance levels. The power rating must be compatible with the load, such as a speaker. Similarly, the impedance levels must be chosen to match the output impedance of the amplification stage to the input impedance of the load for maximum power transfer. If, for example, the output impedance of the amplification stage is 1 kΩ and the speaker impedance is 4 Ω, than an audio transformer with a 1-kΩ primary impedance and 4-Ω secondary impedance is suitable. The frequency response of an audio transformer is defined as the frequency range over which the voltage attenuation is less than 3 dB when terminated with the specified impedance levels. A typical frequency range is 50 to 20,000 Hz. The upper cutoff frequency is limited by the series inductance and shunt capacitance of the windings, and the lower cutoff frequency is limited by the need for increased magnetizing current at lower frequencies resulting from the decrease in primary reactance. Eventually, the voltage drop across the source and primary winding resistance becomes excessive, causing a significant drop in the secondary voltage.

Audio transformers can also be used to couple audio amplification stages, in which case the impedance levels may be, for example, 20 kΩ and 800 Ω. For use with ICs, audio transformers come encapsulated in cylindrical and rectangular shapes with all the wires coming out from one side or from two sides for use with printed circuit boards.

16.4 VIDEO TRANSFORMERS

Video transformers are wide-band devices; that is, their frequency response extends into the megahertz range. Transformers are available with a bandwidth from 20 Hz to several megahertz. The response of the transformer in a system is intimately related to the source and load impedances, and to the wiring, as the frequency-dependent behavior is very sensitive to wiring inductances and capacitances at these high frequencies.

16.5 IF AND RF TRANSFORMERS

IF and RF transformers are used to couple and tune circuits in the hundreds of kilohertz and in the megahertz range. Unlike audio and power transformers, these transformers are not wound on a closed magnetic core; the primary and secondary windings are wound concentrically on a bobbin that contains a movable iron or copper slug inside. The mutual inductance between the coils can be adjusted by lowering or raising the slug; lowering an iron slug into the coils increases the inductance and lowering a copper slug decreases the inductance, since the permeability of copper is less than the permeability of air. Each winding has a capacitor connected across it. Adjustment of the metal core by means of a

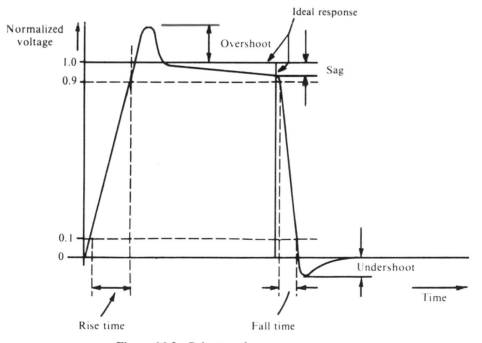

Figure 16-2   Pulse transformer response curve.

screwdriver adjusts the center (peak) frequency. Instead of a variable inductance, the transformer can have screwdriver capacitor adjustments. The primary and secondary circuits can be tuned to the same frequency for sharp and narrow tuning, or to different frequencies for broader band-pass tuning. Similarly, the degree of mutual coupling, by means of the metal slug adjustment, changes the frequency response from a narrow-band single-peak to a wider-band, double-peak characteristic. The fine tuning is done under active operating conditions, that is, when the system is *on*. The important specifications of an IF or RF transformer are the center frequency, the bandwidth range, and the primary and secondary impedance levels.

16.6 PULSE TRANSFORMERS

Pulse transformers are used to couple voltage pulses to change voltage levels, or to isolate electrically one part of a circuit from another. The important specifications of a pulse transformer are, in addition to the voltage and power levels, the *rise time, fall time, overshoot, undershoot,* and *sag* as shown in Figure 16-2.

17
Diodes

17.1 INTRODUCTION

Diodes are the simplest junction semiconductor components. They are used in a variety of applications in which voltages or currents must be rectified. Ideally, a diode does not offer any resistance to current flow in one direction, and blocks completely the current flow in the opposite direction. In practice, this ideal can only be approached. The method by which the ideal diode is approached and the degree to which certain ideal characteristics are approached depends on the purpose for which the diode is manufactured. In this chapter we examine the various diode characteristics and parameters. This is followed with a detailed exploration of diode types.

17.2 DIODE CHARACTERISTICS

Diodes are semiconductor devices that allow current to flow in one direction but not in the opposite direction. Ideally, diodes would have a current-voltage (I-V) characteristic as shown in Figure 17-1. There would be zero voltage drop in the forward direction and zero current when a reverse voltage is applied across the diode—that is, there would be zero leakage current.

Real diodes are different. A typical I-V characteristic is given in Figure 17-2. A finite voltage drop V_D exists in the forward direction, and there is a finite, albeit small, leakage current I_s in the reverse direction. The forward voltage drop at room temperature (25°C) is about 0.7 V in silicon diodes and 0.2 V in germanium diodes. The forward voltage drop decreases with increase in temperature by 2 mV/°C in silicon and 2.5 mV/°C

125

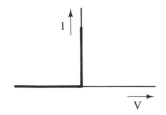

Figure 17-1 Ideal diode voltage-current characteristic.

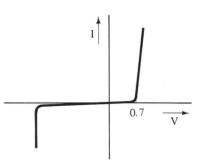

Figure 17-2 Real diode characteristic.

in germanium. The leakage current ranges from nanoamperes in small-current diodes to milliamperes in higher-current diodes. The leakage current doubles in silicon diodes for every 10°C increase in temperature and in germanium diodes for every 6°C increase in temperature.

When a diode is conducting, it offers small but finite resistance of a few ohms or tens of ohms, which for relatively high forward currents causes a voltage drop that may have to be taken into consideration depending on the magnitudes of other voltage drops in the circuit.

Considering these characteristics, the equivalent circuit of a diode is as shown in Figure 17-3. In this figure D is an *ideal* diode, E represents the threshold voltage, R_f is the forward resistance and R_r is the reverse resistance.

When the reverse voltage applied to a diode is increased above a critical value known as the *breakdown voltage*, the diode will conduct current in the reverse direction offering virtually no resistance, causing irreversible destruction of the diode. A diode possesses a ''junction capacitance'' that is inherent in its structure. This capacitance charges up and discharges when the voltage across the diode is reversed. This results in finite switching time, that is, when the voltage across a diode is changed from forward to reverse, the forward current will continue to flow for a short duration. Conversely, when the voltage is switched from reverse to forward, a finite time will elapse before the current reaches steady-state value.

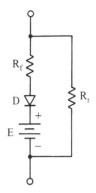

Figure 17-3   Equivalent circuit of real diode.

17.3 DIODE PARAMETERS

The most obvious important parameter is the maximum permissible forward current. This current is specified in terms of the *average current*. Note that when a diode is used to rectify an AC current and it conducts only during half a cycle, the current average over half a cycle can be twice the maximum rated average current. Further, note that for sine waves the peak current is $\pi/2$ (or 1.57) times the average current. Therefore the diode can be used to rectify an alternating current with a peak value equal to $2 \times 1.57 \equiv 3.14$ times the rated maximum average current. In terms of rms values, the diode can rectify a current equal to $(2 \times 1.57)/\sqrt{2} = 2.22$ times the rated maximum average current.

A second important parameter of diodes is the *forward voltage* drop V_D. Since this voltage drop depends on the forward current I_F and the temperature, T, the voltage drop is specified and guaranteed not to exceed a stated value at given current and temperature.

The maximum *peak inverse voltage* (PIV), V_{rp}, is another specified parameter. A reverse voltage exceeding this value causes permanent damage to the diode, which will no longer act as a diode.

Diodes can take a *surge current* that exceeds the maximum average current by up to a factor of 50. The peak surge current is defined as a one-cycle current; usually the pulse width in fractions of a second is also given.

The maximum *leakage current* I_r, or I_s, is also guaranteed in the specification. The temperature and magnitude of the reverse applied voltage for which I_s is measured is also specified.

The maximum *ambient temperature* to which the diode may be subjected is also specified. The value for silicon is typically 125, 150, or 175°C, and for germanium 70°C. The real limiting factor is the junction temperature. Often this temperature is specified. Typical values are 100°C for germanium diodes and 150 to 200°C for silicon diodes. This temperature is related to the ambient temperature through the thermal conductivities of the media and thermal flow barriers through which the heat generated at the junction is dissipated. The mathematical equation expressing these relationships is

$$T_i = P_j(\theta_{jc} + \theta_{cs} + \theta_{sa}) + T_a \tag{17-1}$$

where T_j = junction temperature,
T_a = ambient temperature,
θ_{jc} = junction-to-case thermal resistance,
θ_{cs} = case-to-heatsink (includes insulator) thermal resistance, and
θ_{sa} = heatsink-to-ambient thermal resistance.

The user knows the ambient temperature, and all other data is obtained from manufacturers' data sheets. Some heatsink manufacturers specify the case-to-ambient thermal resistance, θ_{ca}, which combines in series the case-to-heatsink and the heatsink-to-ambient thermal resistances. Equation 17-1 is still applicable if we replace $\theta_{cs} + \theta_{sa}$ with θ_{ca}.

A related value is the maximum permissible power *dissipation* in the diode. The dissipation in watts is equal to the product of the current in amperes and the voltage in volts. Clearly the power dissipated in the diode causes heat flow to the environment, and the junction temperature, which is the ultimate critical factor, is dependent on the power dissipation and heat flow.

There are other diode characteristics that are unique to particular diode types. They will be introduced in the next section.

Equation 17-1 can be used to compute the diode junction temperature for a given power, or to compute the maximum permissible power to keep the junction temperature below the maximum permissible value. But of the most practical value is the problem of determining the required heatsink for the desired application.

Example

Consider a diode that must dissipate 10 W in a given circuit. The maximum allowable junction temperature is 175°C, the ambient temperature is 50°C, and $\theta_{jc} = 2.4$°C/W. Determine the required heatsink.

To conduct sufficient heat away from the device, the thermal resistance θ_{ca} must not exceed a certain value. This value is computed from Equation 17-1 with $\theta_{ca} = \theta_{cs} + \theta_{sa}$. Solving for θ_{ca}, we have

$$\theta_{ca} = \frac{T_i - T_a}{P_j} - \theta_{jc}$$

Substituting numerical values

$$\theta_{ca} = \frac{175 - 60}{10} - 2.4 = 9.1°C/W$$

We use a heatsink listed in a catalog as having $\theta_{ca} = 3.2$°C/W and we know that we are safe. We can compute the actual temperature attained by the diode junction with the chosen heatsink under the stated condition. From Equation 17-1, we have

$$T_j = 10 (2.4 + 3.2) + 60 = 116°C$$

17.4 DIODE TYPES

There are literally hundreds of different diodes. To help users in choosing the right diode, the diodes are divided into several groups. The user can then limit the search, based on the application, to diodes within the appropriate group.

17.4.1 Signal Diodes

Low-power, low-current diodes are often called *signal diodes*. The specifications for the 1N301A silicon diode, for example, are: at 25°C maximum average forward current 65 mA, peak surge current 350 mA for 1 s, maximum dissipation 150 mW, maximum temperature 150°C. Maximum reverse (leakage) current is 0.01 μA with 10-V reverse voltage at 25°C. At 100°C and a reverse voltage of 10 V, the reverse current is 0.20 μA. The diode is guaranteed to allow a minimum current of 18 mA with a forward voltage drop of 1 V. The peak inverse voltage, PIV, is 70 V. The PIV for this group ranges from 1 V to 20 kV, while the majority are rated at tens and hundreds of volts. In addition to the characteristics specified above, many diodes also have the reverse junction capacitance specified. For the 1N301A the value is 5.0 pF at 10-V reverse voltage. The maximum average current in this group is mostly in the range of tens and hundreds of milliamperes. Only very few are rated for 1 A or higher.

17.4.2 Rectifiers

Diodes with higher current ratings are usually classified as rectifiers. Maximum average forward current ratings spread from a few milliamperes to amperes, hundreds of amperes, and even to kiloamperes. PIVs range from 1.5 V to 250 kV. As an example, the 1N2793 silicon diode has the following ratings: PIV 50 V, maximum I_F 8.5 A at 25°C ambient temperature, a one-cycle surge current of 75 A, maximum ambient temperature 175°C, maximum forward voltage drop 1.25 V at 15 A and 150°C case temperature. Maximum leakage current 5.0 mA with 50-V reverse voltage at 150°C.

17.4.3 Switching Diodes/Rectifiers

Switching diodes are characterized by fast switching times. Consider the circuit shown in Figure 17-4. The switch has been in position A. A forward current is flowing through R_L and the diode. At instant $t = 0$ the switch is thrown to position B applying a reverse

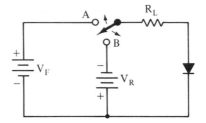

Figure 17-4   Switching diode circuit.

voltage to the circuit. The current flow will not reverse direction instantaneously, but the current will continue to flow in the forward direction for a period t_r, known as the *recovery time*. This is the result of a physics phenomenon at the diode junction known as *minority carrier recombination*. In the fabrication of switching diodes and rectifiers, steps are taken to minimize t_r.

As an example, for the 1N4392 switching diode, the maximum t_r is 500 ps if the forward current before the switching took place was 10 mA, the magnitude of the applied reverse voltage is 6.0 V, and the load resistance, R_L, is 100 Ω. Other specifications of the diode are PIV 15 V, guaranteed minimum current of 2 mA with 1 V forward voltage drop. The leakage current is 1.0 μA at 25°C and 5.0-V reverse voltage. For 5.0-V reverse voltage at 100°C the leakage current is 20 μA. The junction capacitance is 1.0 pF.

The recovery time for switching diodes and rectifiers ranges from 1 ps for low-current diodes to several hundred microseconds for high-current rectifiers.

17.4.4 Reference Diodes

Popularly known as Zener diodes, these diodes are used to regulate voltage or to limit peak voltage. This is accomplished through a controlled reverse voltage breakdown.

At the breakdown voltage the current is limited only by a small dynamic resistance of the diode and other impedance in the circuit. The diode is designed in such a way that even at the breakdown voltage the diode remains intact, provided the power (voltage $\times$ current) dissipated in the Zener diode does not exceed the power rating of the diode.

As an example we list the characteristics of the 1N5222A Zener diode: reference voltage of 2.5 V $\pm$ 10%, maximum dissipation of 500 mW at 25°C, maximum dynamic impedance of 30 Ω at 20 mA. Temperature coefficient is 0.085%/°C; maximum ambient temperature 200°C. The temperature coefficient gives the percentage change in reference voltage per degree Celsius. It is negative for reference voltages below 6 V and positive for reference voltages above 6 V. For the example diode the reference voltage at 50°C is

$$[2.5 \text{ V} - 2.5 \times 0.00085 \times (50 - 25)] \pm 10\% = 2.45 \text{ V} \pm 10\%$$

The range of Zener reference voltages is from 250 mV to 1.5 kV. Common power dissipation ratings are 1, 2, 5, and 10 W, but other ratings from a fraction of a milliwatt to 80 W are available.

18

Component Value Measurement

18.1 INTRODUCTION

Manufacturers mark component values either directly or by code. Usually the values include tolerances. In some applications *precise* values must be used. In this case components are handpicked after measuring their values. In existing equipment the markings are often blurred and hard to read, and when building new circuits it is good practice to check components out before they are soldered in place. In this chapter we present instrumentation techniques for the measurement of component values and characteristics.

18.2 RESISTORS

18.2.1 Ohmmeter

The ohmmeter, or multimeter, is the most common instrument for measuring resistance values. The measurement depends on a voltage ratio, related to the resistor under test and a known reference resistor, not on the absolute value of a voltage. Therefore, slight variation in the battery voltage with aging does not effect the reading. In analog meters the sensing element is a d'Arsonval meter, named after its inventor. This meter consists of a permanent magnet with a moving coil mounted on bearings between the magnet's poles. The force of rotation of the moving mechanism depends on the current through the coil, and the force is balanced by two coiled springs that oppose the rotation of the coil. A pointer attached to the moving mechanism indicates the resistance value on a calibrated scale. Before a reading is taken, the meter must be set to zero. To do this the output terminals of the meter (to which the resistor under test is going to be connected) are shorted, and a variable resistor that is installed in series with the d'Arsonval meter is adjusted to get a zero meter reading. The battery voltage must be high enough to cause sufficient current flow through the resistor under test, but not high enough to cause excessive heating, which would cause resistance changes. The meter has different settings, controlled by a knob, for different resistance ranges. The battery voltage is different for some settings from the values for other settings. It is good practice, if possible, to choose a setting for which the reading is in the center third of the scale, where the reading is most accurate.

In digital meters, an analog-to-digital converter provides a digital output. The meter can be zeroed, with shorted leads, by depressing a key. In autoranging meters, the meter automatically selects the best range.

The accuracy of an ohmmeter depends on the input resistance and other parameters resulting in different accuracies for different ranges. The manufacturer's data sheet must be consulted for greatest accuracy.

18.2.2 Bridge

The accuracy of the simple ohmmeter can be improved by using the d'Arsonval meter only as a null detector. Under this condition the meter provides a reading always when the moving mechanism is in the same position. This is accomplished in the resistance bridge.

The essential parts of a Wheatstone resistance bridge, and modifications thereof (including the Kelvin bridge), consist of two fixed resistors and a variable resistor. The resistor under test forms the fourth arm of the bridge. The resistor under test is connected to the appropriate terminals of the instrument, and the variable resistor is changed until a null reading is obtained, which indicates a balanced bridge. The reading provided by the adjustable resistor setting gives the value of the resistor under test. Several resistance ranges are provided to give maximum accuracy without overheating the resistors. A coarse null adjustment is made first followed by a high-sensitivity adjustment.

The voltage source for the bridge can be DC or AC. The internal AC voltage is usually 1000 Hz. AC voltage makes for a sharper null adjustment, but will include inductance in the measurement. Therefore, wirewound resistors should be read only with a DC source.

Commercial bridges are available with analog and digital readouts, with internal DC and AC (mostly of 1 kHz) excitation and provision for external excitation sources of 20 Hz to 20 kHz and in some cases to 120 kHz. Accuracies are $\pm 1\%$ or $\pm 0.1\%$. Typical resistance ranges are 10 mΩ to 10 MΩ, 1 mΩ to 1.1 MΩ, 1 kΩ to 1,000 TΩ (1 TΩ = 10^{12} Ω), and 50 $\mu\Omega$ to 20 GΩ.

18.2.3 Substitution

In the bridge instrument the accuracy depends, among other variables, on the accuracy of several resistances inside the instrument. By the substitution method, the measurement depends on the accuracy of only one resistor. The bridge is balanced precisely with the unknown resistor. The unknown resistor is then replaced with a precision adjustable resistor. The precision resistor is adjusted to restore a null reading. The setting of the precision adjustable resistor gives the value of the unknown resistor within the accuracy of the bridge.

18.2.4 Megohmmeter

For resistance values above 10 MΩ, high voltages (500 to 1000 V) are necessary to produce sufficient current flow to effect a resistance measurement. A megohmmeter is similar to the ohmmeter discussed in Section 18.2.1, except that it contains a high supply voltage and that the d'Arsonval meter is provided with a very high input resistance. This instrument is capable of resistance measurement to 1 TΩ (10^{12} Ω). Remember that moisture and certain contaminants on the surface of the resistor can contribute to the conductance, which can be significant for high resistances. Therefore, a high-value resistor must be cleaned with a solvent before measurements are made.

18.3. CAPACITORS

The important parameters of real (rather than ideal) capacitors are defined and discussed in Chapter 14. Refer to these definitions if needed.

18.3.1 Insulation-Resistance (DC-Leakage) Measurement

The direct way to measure this resistance is to apply a DC voltage to the capacitor and measure the current. This is difficult to do with high-quality capacitors. Consider, for example, a 1-μF polycarbonate capacitor whose megohm-microfarad product is 50,000 s. Its computed insulation resistance is 50,000 MΩ. The DC voltage that can be applied to the resistance measurement is limited by the capacitor voltage rating. Suppose that the voltage is 500 V. The measured current will be 500 V/50,000 MΩ = 10 nA. A very sensitive ammeter is required. This method can conveniently be used with electrolytic capacitors which have relatively high leakage current. Since the insulation resistance is dependent on the voltage, the leakage current must be measured with a voltage equal to the voltage to which the capacitor will be subjected in the circuit.

   An indirect way to measure the insulation resistance is by way of the discharge time constant. The capacitor is charged to the operating voltage, V_O. The voltage is measured after a time interval ΔT, and the difference between the original voltage and the present voltage is designated ΔV. The time constant, TC, is computed as

$$TC = V_O \frac{\Delta T}{\Delta V} \qquad (18\text{-}1)$$

The time constant is equal to the product of the insulation resistance R_I and the capacitance C.

$$TC = R_I \times C \qquad (18\text{-}2)$$

Combining Equations 18-1 and 18-2, we have

$$R_I = \frac{V_O \, \Delta T}{C \, \Delta V}$$

If the capacitance is known, R_I can be computed. We show in the next section how to measure the capacitance.

The time interval ΔT can be seconds, minutes, or hours. It depends on the capacitor quality and is chosen so that a small change in voltage can be measured easily (e.g., $\Delta V = 3$ or 5% of V_O). If ΔV is too large, the result is inaccurate, since $\Delta V / \Delta T$ must represent the angle of the tangent to the exponential discharge curve at V_O. To make sure that the voltmeter does not contribute significantly to the capacitor discharge, the voltmeter must have input impedance higher than the discharge resistance and must be connected to the capacitor only momentarily.

18.3.2 Bridges—Capacitance and Dissipation Factor Measurement

In Chapter 14, we expressed the capacitor including the dissipation factor in terms of a series resistance-capacitance network. Alternatively, we can express the capacitor in terms of an equivalent parallel resistance-capacitance network. The equivalency of the two representations is shown in Figure 18-1.

The capacitance measurement is performed with the help of a capacitance bridge. The unknown capacitor is connected to two terminals, and the general procedure is similar

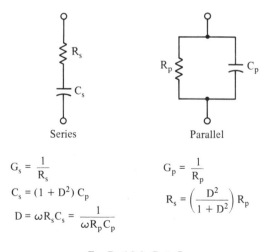

Series   Parallel

$$G_s = \frac{1}{R_s} \qquad\qquad G_p = \frac{1}{R_p}$$

$$C_s = (1 + D^2)\,C_p \qquad R_s = \left(\frac{D^2}{1 + D^2}\right) R_p$$

$$D = \omega R_s C_s = \frac{1}{\omega R_p C_p}$$

For $D < 0.1$, $C_s \simeq C_p$

Figure 18-1   Equivalent circuits for capacitor, C, including the dissipation factor, D.

to the measurement of a resistor with a resistance bridge, except that two adjustments are necessary. In Figure 18-1 we note that there is a real part and an imaginary part. Both parts must be balanced by the bridge using the null detector. In general, several iterations are necessary to balance both parts.

Different bridges give different readout combinations. For example, one type gives readouts of C_p and D, another of C_s and D, another of C_p and G_p, and still another of C_p and R_p. Using the relationships given in Figure 18-1, one can convert measured parameters to another set of parameters to derive an equivalent circuit most useful for the application.

Note that all the parameters are functions of frequency, so that the values have meanings only if measured at the applicable frequency. Most commercial capacitance bridges have internal voltage sources as well as provisions for external sources. Common values are 1 kHz internal source and provisions for 20-Hz to 20-kHz external source. However, there are bridges with which 300-kHz voltage sources can be used, and there are bridges (particularly for small capacitance values) with internal sources of 1 MHz and internally adjustable frequencies from 1 MHz to 100 MHz. Examples of capacitance ranges of capacitance bridges are 1 pF to 1000 μF, 0.01 pF to 1100 μF, 2.0 pF to 1.1 F, 0.02 pF to 15 pF, and 0.00005 pF to 1000 pF. Accuracies range from 1% to 0.1%.

In high-frequency bridges parasitic effects become important. Special means are provided to cancel these parasitic effects and improve resolution.

18.4 INDUCTORS

Real inductors are lossy—that is, they have parasitic resistance that dissipates energy. This resistance degrades the performance of the inductor and must be considered in critical designs. This property is most readily expressed in terms of the quality factor, Q, of the inductor. The resistance may be detrimental at one frequency, but not at another. The definition of Q and its significance are given below. There are two equivalent circuits for real inductors: the RL series equivalent circuit and the RL parallel equivalent circuit. The circuits are shown in Figure 18-2 together with the relevant equations. Note that using these equations allows the parameters of either equivalent circuit to be computed from the known parameter values for the other equivalent circuit.

The quality factor Q is defined as the ratio of the inductive reactance at the relevant frequency to the series resistance of the series equivalent circuit. The higher the Q, the less lossy and the more nearly ideal is the inductor. In the parallel representation the equivalent resistance shunts the inductance, and therefore the larger the resistance, the more nearly ideal is the inductor, again corresponding to a large Q. Note that Q is frequency-dependent.

18.4.1 Bridges

As in the capacitance bridge, the inductance bridge must null two parts: the real and the imaginary parts of the unknown impedance. Therefore two adjustments, usually iter-

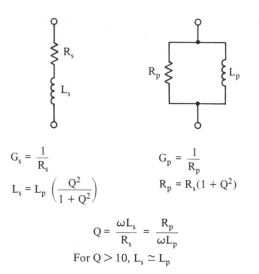

$$G_s = \frac{1}{R_s} \qquad\qquad G_p = \frac{1}{R_p}$$

$$L_s = L_p \left(\frac{Q^2}{1 + Q^2}\right) \qquad\qquad R_p = R_s(1 + Q^2)$$

$$Q = \frac{\omega L_s}{R_s} = \frac{R_p}{\omega L_p}$$

For $Q > 10$, $L_s \simeq L_p$

Figure 18-2 Equivalent circuit for inductor.

atively, must be made. The impedance can be expressed in terms of the series or parallel inductance and the corresponding resistance or conductance, or the quality factors. The relationships among these quantities are given in Figure 18-2. Whichever parameters the bridge provides, the user can convert them to the quantities most directly applicable to the particular design.

Some commercial bridges provide readouts of L_s and R_s, others provide readouts of L_p and G_p, or L_s and Q, or L_p and Q. Inductance ranges of inductance bridges include 1 µH to 1000 H, 0.05 µH to 1100 H, 2 µH to 11 mH, 0.1 nH to 1,111 H, and other ranges. Accuracies range from 0.1% to 1%. Some bridges contain internal voltage sources for measurements, as well as provisions for external sources; others have only internal sources or only a provision for external sources. The designs can accommodate frequencies from 1 kHz to 500 kHz.

18.4.2 Q Meters

High values of Q can be measured with high accuracy using an instrument known as the Q meter. This instrument consists of a low-output-impedance sine-wave generator and a high-input-impedance voltmeter. The unknown inductor is connected in series with a variable precision capacitor. The voltmeter, V_2, is connected across the capacitor. The signal generator is adjusted to the desired frequency, f, and the capacitance is then varied until it resonates with the unknown inductance indicated by a sharp peak in the voltmeter reading, V_2. At resonance, $L_s = \frac{1}{2\pi f C}$. Since f and C are known from dial settings, L_s can be computed.

The voltage, V_1, at the terminals of the signal generator is also read. At resonance, $V_2/V_1 = \omega L_s/R_s = Q$, so that knowing L_s, R_s can also be computed.

18.5 DIODES

18.5.1 Curve Tracer

Most conveniently the diode characteristics are measured with a curve tracer. The diode is connected to the instrument terminals where it is put in series with a selectable resistor. A varying voltage is applied to the circuit, and the *I-V* characteristic is displayed on a calibrated CRT. The leakage current, forward voltage drop, Zener break-down voltage (for a reference diode), and the detailed *I-V* relationship can also be read off the CRT.

18.5.2 Volt-Ampere Method

A less-sophisticated method consists of connecting in series a DC variable voltage source, the diode, a current limiting resistor, and an ammeter (microammeter, nanoammeter, or picoammeter, as the need may be). A voltmeter is connected across the diode. When the voltage source applies a forward voltage to the diode, the forward voltage drop at a given current is measured. To measure the leakage current at a reverse voltage, the voltage source polarity connections are reversed to apply a reverse voltage to the diode. The voltages are read across the diode and the resistor. The latter reading gives the current by dividing the measured voltage by the resistance value. This gives the leakage current as a function of reverse voltage.

19

Component Identification

19.1 INTRODUCTION

Often components are too small to print on them legible numbers for value identification. Manufacturers use color codes to verify component values and tolerances. Once one is familiar with the color code, the code actually expedites practical experimentation and assembly work. For example, if one needs a resistor in the kilohm range, one looks for a resistor with a red third band. Resistors in the hundreds of kilohms range have a yellow third band. The fourth band identifies the tolerance. In this chapter you will find the various codes and designations used to identify information pertaining to resistors, capacitors, diodes, and transformers.

19.2 RESISTORS AND POTENTIOMETERS

The resistance value of carbon resistors is signified by color bands. There are either three bands or four bands. The group of bands is closer to one end of the resistor than the other. Beginning with the first band near the closest end, the first *three* bands have the following meaning. The first two bands represent the first two digits of the resistance value, and the third band gives a multiplier in terms of the power of ten by which the number formed by the first two digits must be multiplied.

The color code is as follows:

Black	0	Orange	3
Brown	1	Yellow	4
Red	2	Green	5

Blue	6	White	9
Violet	7	Gold	−1
Gray	8	Silver	−2

(*Note:* In a three-band code only the third band can be gold or silver.)

Examples

Red, violet, orange

$$27 \times 10^3 = 27,000 \ \Omega = 27 \ k\Omega$$

Blue, orange, gold

$$63 \times 10^{-1} = 6.3 \ \Omega$$

The tolerance of three-band coded resistors is ±20%.

For tighter tolerances, a four-band code is used. The fourth band is either silver for ±10% tolerance or gold for ±5%.

The power rating of carbon-composition resistors are recognized by the physical dimension: a ⅛-W resistor is 0.145 in. long and has a 0.062-in. diameter. For a ¼-W resistor the dimensions are 0.25 and 0.090 in.; for a ½-W resistor, 0.375 and 9⁄64 in.; for a 1-W resistor, 9⁄16 and 7⁄32 in.; and for a 2-W resistor, 11⁄16 and 5⁄16 in.

Precision wirewound resistors and higher-power resistors usually have their resistance values and power ratings printed directly on their body.

Potentiometers often have the resistance and power values stamped directly on their casings. Occasionally the resistance value is given by code. The code is similar to the "band" code, except that instead of colors, digits are used.

Example

A potentiometer with the number 103 stamped on it has a resistance
$10 \times 10^3 = 10 \ k\Omega$
The number 504 represents $50 \times 10^4 = 500 \ k\Omega$.

19.3 CAPACITORS

Capacitor markings are more confusing than resistor markings and in general require a reference table.

19.3.1 Electrolytic Capacitors

These capacitors are usually cylindrical and have axial leads. The capacitance values and voltage ratings, as well as polarity, are written directly on the case. The capacitance values are in microfarads. Often the unit is indicated as ''µF.'' In some cases an incorrect and misleading notation is used, namely ''mF.'' The intention is still to indicate microfarads.

19.3.2 Paper Dielectric Capacitors

These are tubular in form with the values and ratings written directly on them. In addition there is a black band at one end. The terminal near the band is connected to the outer foil of the capacitor and should be connected to the lower voltage to minimize electric noise pickup.

19.3.3 Ceramic Capacitors

The capacitors are in a shape of either a disk or a cylinder. The disk types usually have the capacitance values and voltage ratings written directly on them. The capacitance *units* are often not indicated. Whole numbers are in picofarads (e.g., 33 pF), while decimal fractions are in microfarads (e.g., 0.05 µF). Sometimes a three-digit code is used for the capacitance value. This is similar to the code used for resistors. For example, ''223'' means ''22×10^3'' or 22,000 pF. Sometimes a color code is used in the form of dots or stripes. The dots or stripes representing the first two digits use the same color code as resistors. The third-digit color code (i.e., the power of 10) differs: black (0), brown (1), red (2), orange (3), yellow (4), gray (-2), white (-1).

When digits are used, tolerances are given by letter: for capacitance greater than 10 pF, ±20% is M, ±1% is F, ±2% is G, ±3% is H, ±5% is J, and ±10% is K. The corresponding color codes are black, brown, red, orange, green, and white.

For capacitance less than 10 pF the tolerances are ±2.0%, G; ±0.1%, B; ±0.5% D; ±0.25%, C; ±1.0%, F. The corresponding color codes are black, brown, green, gray, and white.

The color coding for temperature coefficients of capacitance follows the color sequence (from 0 through 9) given in Section 19.2. The code is given in ppm/°C (parts per million per degree Celsius): 0, -33, -75, -150, -220, -330, -470, -750, -1500 to $+150$, -750 to $+100$. The positioning of the color codes is shown in Figure 19-1.

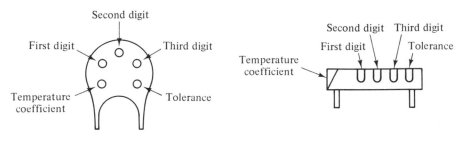

Figure 19-1 Ceramic capacitors.

19.3.4 Mica Capacitors

Mica capacitors are low in capacitance values and often carry the value written directly with the "pF" notation. (Occasionally a misleading and incorrect notation "mmF" is used. The intention is "$\mu\mu$F," which is the same as "pF.")

Sometimes capacitance is given in symbolic digits, and the tolerance is given by a letter. The first two digits represent a number, and the third digit represents a power of 10. Units are picofarads. For example, "223" means $22 \times 10^3 = 22,000$ pF. The percentage tolerance letter coding is ± 20, M; ± 1, F; ± 2, G; ± 5, J; ± 0.5, G; ± 10, K.

Color codes corresponding to these tolerances are, respectively, black, brown, red, green, gold, silver. Temperature coefficients (ppm/°C) also have color codes: brown, ± 500; red, ± 200; orange, ± 100; yellow, -20 to $+100$; green, 0 to $+70$.

Color codes for capacitance values use the same numbering system as resistors (Section 19.2). The units are picofarads.

The positioning of the color code is shown in Figure 19-2.

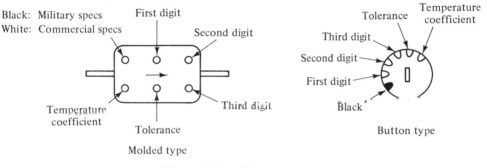

Figure 19-2   Mica capacitors.

19.4 INDUCTORS

The inductance value is usually listed directly on the case. Very small inductors, in the microhenry range, are usually packaged in cylinders and look similar to resistors. The identification can be done with an ohmmeter, as an inductor has very low DC resistance and a capacitor is an open circuit to DC.

19.5 TRANSFORMERS

Often transformers have the voltages and power rating in VA stamped right on the metal frame or on the encapsulation. Sometimes only the model number is given, in which case the manufacturer's catalog must be consulted. The standard color code used for transformer wires is illustrated in Table 7 of Appendix B.

19.6 DIODES

Small-size diodes are cylindrical and look similar in shape to a ¼-W resistor. The diode body is usually black (whereas ¼-W resistors are brown). Diodes are identified by the prefix "1N" and a number. The number can have two, three, or four digits (e.g., "1N4148"). Sometimes instead of numerical digits, color bands are used. In such cases the prefix "1N" is understood (not printed) and the color code is identical to that of resistors (Section 19.2). Thus "yellow, brown, yellow, gray" stands for "1N4148."

The cathode and anode of the diode are often identified by the diode symbol printed on the diode:

Cathode ———◄|——— Anode

Sometimes the color code band's position is used to identify the anode and cathode. The set of color bands is closer to one end than the other. The closer end is the cathode. The band closest to the cathode is the first digit. In physically small diodes the color bands cover the whole diode body. In this case the color band closest to the cathode is made wider than the other bands. The wider band also represents the first digit.

Part IV PUTTING IT ALL TOGETHER

20

Comprehensive Design Examples

20.1 INTRODUCTION

In this chapter we develop in complete detail two design examples. First we design a band-pass filter, then a low-pass filter. In addition to deciding on the correct circuits and selecting all the components, including values and specific materials, we also show the effect of component tolerances and temperature on the filter characteristics.

20.2 DESIGN 1—BAND-PASS FILTER

A signal voltage in a system operated at room temperature must be filtered through a band-pass filter with a center frequency, f_c, of 200 kHz. To attenuate voltages of un-wanted frequencies, the quality factor, Q, of the filter must equal 5. The 200-kHz voltage must be amplified by a factor of 50. The system DC supply voltages are ± 12 V.

Design the filter.

20.2.1 Solution

Searching for a suitable circuit configuration we find that the filter can be implemented with the circuit shown in Part I, Figure 3-3, using Table 3-3. We must (1) select an appropriate operational amplifier, (2) compute passive component values, and (3) choose available components.

Operational amplifier—selection

The center frequency of 200 kHz rules out the use of the popular LM 741 internally compensated amplifier. The active filter provides gain at the desired frequency and we must allow for 10, 20, or more decibels. The LM 741 opamp limits the gain at 200 kHz to less than 15 dB (see Appendix A1 and Part II). The uncompensated opamp model MC1439P1 is suitable. This amplifier is similar to the MC1539G shown in Part II. The latter has an operating temperature range from -55 to $+125°C$ and is packaged in a metal can, while the former has an operating temperature range from 0 to $+70°C$ and is packaged in a plastic DIP (dual-in-line package). Since our system requires only room temperature operation, the less expensive MC1439P1 is the correct choice.

From pp. 78, we note that to compensate the opamp for our application, we must connect in series a 1-kΩ resistor and a 2200-pF capacitor between the compensating terminals, terminals 1 and 8.

From Table 3-3 we choose for the filter an amplifier gain of $K = 3.71716$.

Component values computation

R_1, R_2, R_3, C_1, C_2 (Figure 3-3).   Table 3-3 gives for $K = 3.71716$, $Q = 5$, and angular center frequency $\omega_c = 1$, the following values:

$$R_1 = R_2 = R_3 = 1.41421 \ \Omega$$
$$C_1 = C_2 = 1 \ F$$

To shift the center frequency to $f_c = 200$ kHz, we compute

$$\omega_c = 2\pi f_c \times 200,000 = 1,256,637 \ \text{rad/s}$$

We divide the values of the capacitors by 1,256,000. This gives

$$C_1 = C_2 = 0.79517 \ \mu F$$

To obtain realistic resistance values, we must multiply the resistance values given above by some factor. In order not to change the center frequency, we must divide the capacitance values by the same factor. An upper limit for resistance values is dictated by the amplifier bias current of the amplifier, which must be much smaller than the current through the resistors. The bias current is specified on the data sheets as 200 nA. The current, i, through a resistor can be at most the power supply voltage divided by the resistance. It follows that for any resistor we must have $R \ll (12 \ V/200 \ nA) = 600 \ \Omega$. Accordingly, we multiply all resistance values by 7.0710856 and divide the capacitance values by the same factor. (The particular value 7.0710856 was chosen to give the resistors a convenient round value in the correct range.) Thus,

$$R_1 = R_2 = R_3 = 10 \ k\Omega$$
$$C_1 = C_2 = 112.5392 \ pF$$

To give the amplifier a gain of $K = 3.71716$, we pick

$$R_a = 10 \text{ k}\Omega \text{ and } R_b = 27,172 \text{ k}\Omega$$

Finally, we use power supply decoupling capacitors: 0.1 μF ±20% ceramic disks, rated at 25 VDC. They are about 0.6 in. in diameter. If the large size is a problem, high-K ceramic capacitors can be used with a size of about 0.2×0.2 in., at the expense of somewhat higher dissipation factor.

Material selection—specific components

Resistors.   The value of f_c is critically dependent on the RC product. We pick a standard commercially available R value close to the computed value. We then pick a standard commercially available C value, but shunt each capacitor with a trimmer so that we can tune the RC product to the exact value required.

To determine the required power ratings of the resistors, one can compute the power dissipated in each resistor, but the power requirements are so small that it is not necessary to make such detailed computation. An upper limit of dissipated power is obtained by taking the highest voltage to which any component in the circuit can be subjected and using this value to compute the power in the smallest resistor.

$$(V^2/R) = (12^2/10,000) = 0.0144 \text{ W}$$

Hence ⅛-W resistors are sufficient.

Note that although the exact value of R is not important (since the RC product can be fine tuned with the trimmer capacitors), it is important that all the resistors, R_1, R_2, and R_3, have the same value R. Therefore we shall use 10 kΩ carbon-film resistors, which can be purchased with ±0.5% tolerance.

To obtain the exact value for the gain $K = 3.71716$ we use one fixed resistor and one potentiometer. Since the potentiometer will be used to adjust for the exact gain, the fixed resistor $R_4 = 10 \, K$ can be an inexpensive ±10%-tolerance ⅛-W carbon composition resistor. For the potentiometer we use a ¼-W 50-K cermet unit.

Capacitors.   The system is operating at relatively high frequency. Therefore we choose ceramic capacitors that maintain their capacitance at high frequency (see Part III).

Reference to manufacturer catalogs shows that a suitable ceramic capacitor close to the value we computed is 100 ±5 pF, 1000 VDC, 0.25 in. in diameter. To allow for fine tuning, we parallel the capacitors with capacitor trimmers adjustable between 3.5 pF and 20 pF, 100 VDC.

The schematic diagram of the filter is shown in Figure 20-1, and the bill of materials is given in Table 20-1.

The properly tuned filter has a frequency response as shown in Figure 20-2. Note that the frequency is plotted on a logarithmic scale. In fact, the frequency response is not symmetric with respect to the center frequency. To illustrate this, we show a frequency response curve of the same filter with the frequency plotted on a linear scale in Figure 20-3.

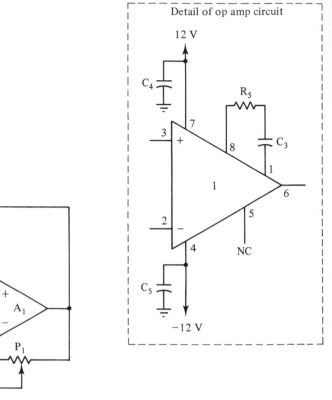

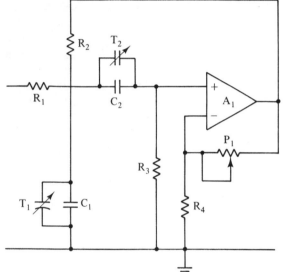

Figure 20-1 Band-pass filter with f_c = 200 kHz.

TABLE 20-1 Bill of Materials for Circuit of Figure 20-1

Item	Symbol	Description	Quantity
1	A_1	OPAMP, MC1439PI	1
2	R_5	Resistor, 1 kΩ ±0.5%, 1/8 w, carbon film	1
3	$R_1 R_2 R_3$	Resistor, 10 K ±0.5%, 1/8 w, carbon film	3
4	R_4	Resistor, 10 K ±10%, 1/8 w, carbon composition	1
5	P_1	Potentiometer, 50 K, 1/4 W, cermet	1
6	$C_1 C_2$	Capacitor, 100 pF ±5%, 1000 VDC, ceramic	2
7	C_3	Capacitor, 2200 ±10%, 1000 VDC, ceramic	1
8	C_4, C_5	Capacitor, 0.1 μF ±20%, 25 VDC, ceramic	2
9	T_1, T_2	Capacitor trimmer, 3.5 pF − 20 pF, 100 VDC	2

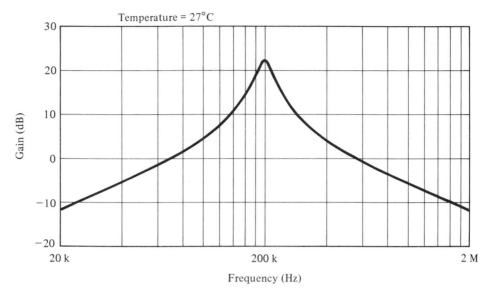

Figure 20-2   Frequency response of filter of Figure 20-1.

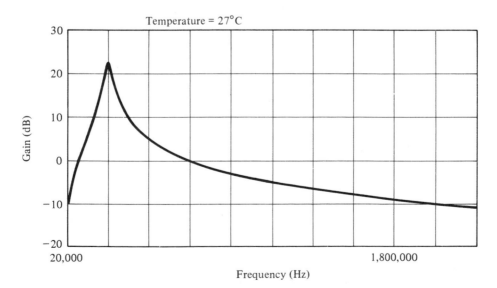

Figure 20-3   Frequency response of filter of Figure 20-1—linear frequency scale.

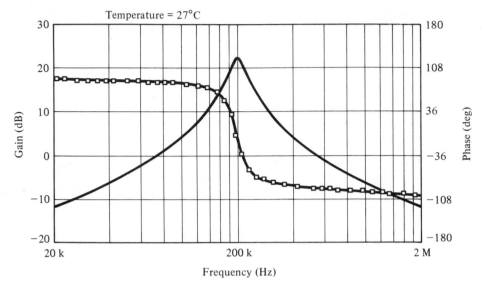

Figure 20-4   Amplitude and phase frequency response of filter of Figure 20-1.

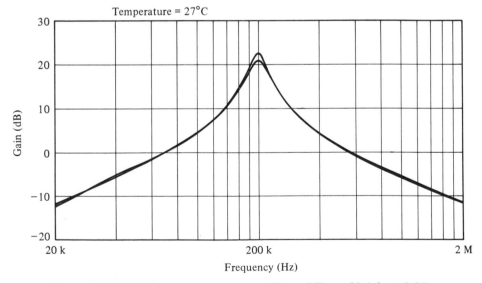

Figure 20-5   Frequency response curves of filter of Figure 20-1 for ±0.5%-tolerance resistors.

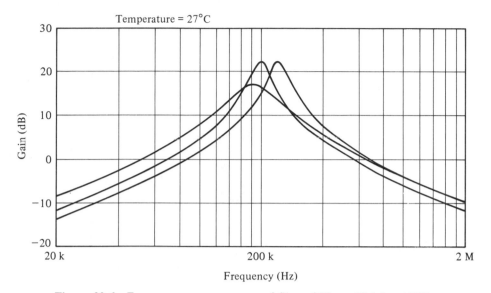

Figure 20-6 Frequency response curves of filter of Figure 20-1 for ±20%-tolerance resistors.

In Figure 20-4 we show the phase-shift together with the amplitude frequency response. The phase shift is zero at the center frequency.

The importance of using high precision resistors in the design of this filter is apparent from Figures 20-5 and 20-6. Figure 20-5 shows three frequency response curves of the filter constructed with 0.5% tolerance resistors. The curves were obtained for three sets of randomly selected resistors within the tolerance. Note that two curves almost coincide, and the third curve is only slightly different. By contrast, Figure 20-6 shows three frequency response curves for the same filter with the same nominal resistor values, but with ±20% tolerance. Again, the curves were obtained using three sets of randomly selected resistors within the tolerance value. The wide spread in frequency response is readily seen.

To obtain the gain of 50 specified in the design example statement, we cascade the filter with an amplifier of adjustable gain as shown in Figure 20-7. The gain of the filter without the additional amplification stage is at the center frequency about 22.5 dB or 13.34 (see Figure 20-2). The amplifier stage must provide a gain of (50/13.34) = 3.75, giving a total gain of 34 dB. For this stage we use the same opamp type that we used for the filter, type MC1439P1 with the same compensation network. We use the amplifier in the noninverting mode with a negative feedback network consisting of a 50-kΩ cermet potentiometer and a $10K$ ± 10% carbon-composition resistor to establish the required gain. The complete bill of materials is given in Table 20-2. The frequency response of the overall system is shown in Figure 20-8.

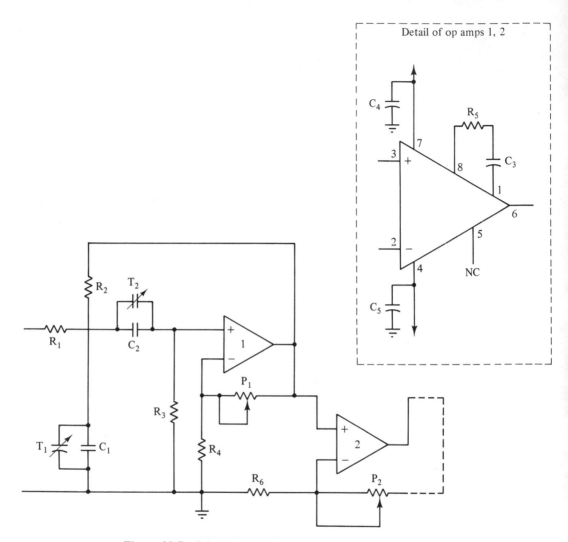

Figure 20-7 Schematic circuit diagram of filter cascaded with amplifier.

TABLE 20-2 Bill of Materials for Circuit of Figure 20-7. This Table is Same as Table 20-1, Except Quantities Differ as Follows:

Item	Quantity
1	2
2	2
4	2
5	2
8	4

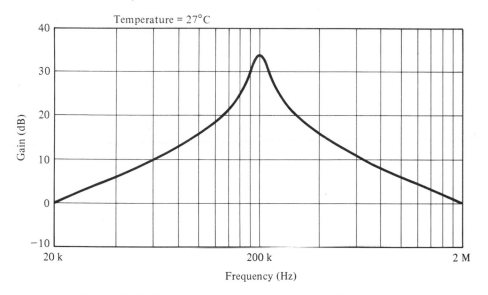

Figure 20-8   Frequency response of the system of Figure 20-7.

20.3 DESIGN 2—LOW-PASS FILTER

Design a low-pass filter with a high-frequency cutoff $f_H = 4$ kHz $\pm$ 10%. The frequency response must be flat in the passband and attenuate sharply, at a rate in excess of -100 dB/dec, in the stop band. The filter is to be used at the front end of a system where the signal level is low. The supply voltage is ± 9 V. The signal source has an output resistance in the megohm range. The system must function in the temperature range of -25 to $+85°C$.

20.3.1 Solution

The requirements of flat response in the passband and attenuation in excess of -100 dB/dec in the stop band can be met with a maximally flat, sixth-order filter (Part I, Chapter 6). We must (1) select the appropriate operational amplifiers, (2) compute passive component values, and (3) choose available components made of suitable materials.

Operational amplifier selection

The signal source has a high output impedance and a low signal level. This calls for a low-noise amplifier with a very high input impedance. Studying the data sheets given in Appendix A3, we find that the precision LH0052 FET opamp meets the requirements. It has an input resistance of 12 MΩ, the input offset current is less than 100 fA, the offset voltage is 200 μV, the input noise current is less than 1 pA over a bandwidth from 10 Hz to 10 kHz, and the input noise voltage is in the microvolt range. The amplifier is internally compensated, has DC gain of 180,000, and unity gain bandwidth of 1 MHz suitable for a filter with a 4-kHz cutoff frequency.

Component values computations

The component values for a low-pass sixth-order filter with a 4-kHz cutoff frequency were computed in Part I, Chapter 6. The resistor values are: 682 Ω, 5,858 kΩ, 10 kΩ, and 14.824 kΩ. All capacitors have a value of 3,979 pF.

Material selection—specific components

Resistors.   Since this design emphasizes low noise and a wide temperature range, we decide to use *metal-film resistors*. They are available with a temperature coefficient of resistance (*TCR*) as low as 1 ppm/°C, and they have extremely low noise figures (see Part III).

Standard metal-film resistor values close to those we calculated are 681 Ω, 5.9 kΩ, 10 kΩ, and 14.7 kΩ.

Power ratings.   The highest voltage that can appear across a resistor is 9 V. The power dissipated in a resistor is given by the voltage squared divided by the resistance. In the resistors in the filter circuit the power dissipation is less than 0.01 W (note that the 681-Ω and 5.9-kΩ resistors are each in series with a 10-kΩ resistor, resulting in reduced voltages across them). From the power derating equation in Part III, we have at 85°C

$$100\% \text{ power rating } [1 - (85 - 70)/80] = 100\% \times 0.8125$$

For a ⅛-W resistor, the power rating at 85° is $0.125 \times 0.8125 = 0.1016$ W. Hence a ⅛-W resistor is a good choice.

Capacitors.   Pure reactance is noiseless. Capacitors, however, have parasitic resistance that is a source of noise. This resistance is expressed as a dissipation factor (or its reciprocal, the *Q* factor). The lower the dissipation factor, the lower the resistance and the lower the noise.

A study of Table 14-1, Part III, reveals that polystyrene capacitors are an excellent choice for our application. Their dissipation factor is only a few hundredths of a percent throughout the applicable temperature range, and their capacitance changes by less than 1% over the temperature range.

A commercially available polystyrene capacitor close in value to the computed value is 3900 pF, 33 VDC.

Finally, we include in the circuit power supply decoupling capacitors: 0.1-μF ceramic disks.

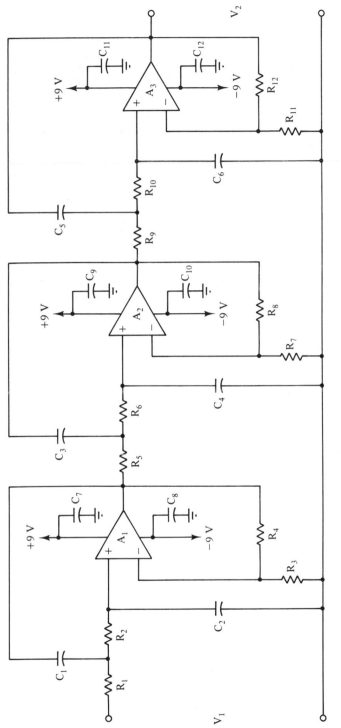

Figure 20-9 Sixth-order low-pass filter.

TABLE 20-3   Bill of Materials for Circuit of Figure 20-9

Item	Symbol	Description	Quantity
1	$A_1 A_2 A_3$	OPAMP, FET, LH0052	3
2	R_4	Resistor, 681 ±1%, $^1/_8$ W, metal film	1
3	R_8	Resistor 5.9 K ±1%, $^1/_8$ w, metal film	1
4	$R_1 R_2 R_3 R_5 R_6 R_7$ $R_9 R_{10} R_{11}$	Resistor 10 K ±1%, $^1/_8$ w, metal film	9
5	R_{12}	Resistor 14.7 K ±1%, $^1/_8$ w, metal film	1
6	$C_1 C_2 C_3 C_4 C_5 C_6$	Capacitor 3900 pF ±5%, 33 VDC, polystyrene	6
7	$C_7 C_8 C_9 C_{10} C_{11} C_{12}$	Capacitor 0.1 μF ±10%, 25 VDC, ceramic disks	6

Comments

The nominal capacitance value of 3900 pF is 2% lower than the computed value. The tolerance is ±5%. The resistor values are within 1% of the computed values. Therefore, we are within the required 10% tolerance of the cutoff frequency.

The schematic circuit diagram of the filter is shown in Figure 20-9. The bill of materials is given in Table 20-3. Figure 20-10 shows the frequency response of the filter at 25°C with nominal component values. The spread of the response over the temperature range from −25 to +85°C is shown in Figure 20-11. This figure also includes responses for randomly picked component values within the specification tolerances.

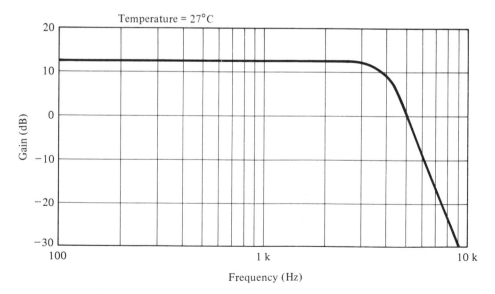

Figure 20-10   Frequency response of filter of Figure 20-9 at 25°C with nominal component values.

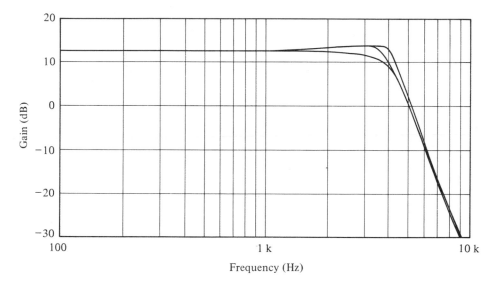

Figure 20-11   Spread of frequency response of filter of Figure 20-9 between temperature limits of −25 and 85°C including component value tolerance.

20.4 SUMMARY AND COMMENTS

In this part of the book we have shown how to use information in other parts by referring to different sections and combining material in a way that results in practical design of filters. This includes circuitry and specification of components. The drawings and bills of material are complete and ready to be given to the fabricator for implementation.

The designer must have access to data sheets of operational amplifiers, resistors, and capacitors. Selected data are given in this design book. To be informed of all relevant material, the designer must keep updated files of data sheets from manufacturers, distributors, and suppliers of electronic components.

APPENDIX A
Manufacturers' Data Sheets

Appendix A contains data sheets provided by manufacturers. The selection of data sheets is representative of the various operational amplifier types.

1. Internally compensated operational amplifier
2. Uncompensated operational amplifier
3. Very-high-input-impedance operational amplifier
4. Transconductance operational amplifier
5. Single-polarity-power-supply operational amplifier
6. Low-voltage-power-supply operational amplifier
7. Very-low-voltage-power-supply operational amplifier
8. Low-noise operational amplifier
9. Wideband operational amplifier

1. INTERNALLY COMPENSATED OPERATIONAL AMPLIFIER

(Reprinted by permission of National Semiconductor Corp.)

Operational Amplifiers

LM747/LM747C dual operational amplifier[*]

general description

The LM747 and the LM747C are general purpose dual operational amplifiers. The two amplifiers share a common bias network and power supply leads. Otherwise, their operation is completely independent.

features

- No frequency compensation required
- Short-circuit protection
- Wide common-mode and differential voltage ranges

- Low-power consumption
- No latch-up
- Balanced offset null

Additional features of the LM747 and LM747C are: no latch-up when input common mode range is exceeded, freedom from oscillations, and package flexibility.

The LM747C is identical to the LM747 except that the LM747C has its specifications guaranteed over the temperature range from 0°C to 70°C instead of −55°C to +125°C.

schematic diagram (each amplifier)

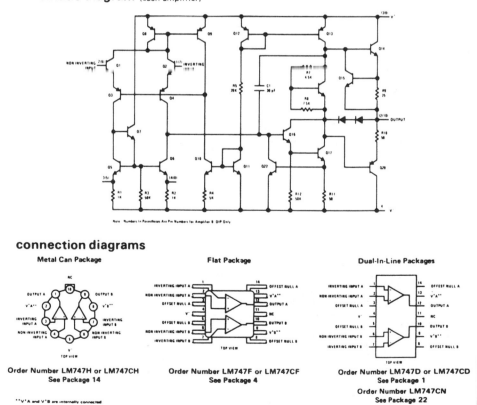

Note: Numbers In Parentheses Are Pin Numbers for Amplifier B DIP Only

connection diagrams

Metal Can Package

Order Number LM747H or LM747CH
See Package 14

**V⁺A and V⁺B are internally connected

Flat Package

Order Number LM747F or LM747CF
See Package 4

Dual-In-Line Packages

Order Number LM747D or LM747CD
See Package 1

Order Number LM747CN
See Package 22

[*] The 741 amplifier has identical characteristics.

absolute maximum ratings

Supply Voltage LM747	±22V
LM747C	±18V
Power Dissipation (Note 1)	800 mW
Differential Input Voltage	±30V
Input Voltage (Note 2)	±15V
Output Short-Circuit Duration	Indefinite
Operating Temperature Range LM747	$-55°C$ to $125°C$
LM747C	$0°C$ to $70°C$
Storage Temperature Range	$-65°C$ to $150°C$
Lead Temperature (Soldering, 10 sec)	$300°C$

electrical characteristics (Note 3)

PARAMETER	CONDITIONS	LM747			LM747C			UNITS
		MIN	TYP	MAX	MIN	TYP	MAX	
Input Offset Voltage	$T_A = 25°C$, $R_S \leq 10 \text{ k}\Omega$		1.0	5.0		1.0	6.0	mV
Input Offset Current	$T_A = 25°C$		80	200		80	200	nA
Input Bias Current	$T_A = 25°C$		200	500		200	500	nA
Input Resistance	$T_A = 25°C$	0.3	1.0		0.3	1.0		$M\Omega$
Supply Current Both Amplifiers	$T_A = 25°C$, $V_S = \pm15V$		3.0	5.6		3.0	5.6	mA
Large Signal Voltage Gain	$T_A = 25°C$, $V_S = \pm15V$ $V_{OUT} = \pm10V$, $R_L \geq 2 \text{ k}\Omega$	50	160		50	160		V/mV
Input Offset Voltage	$R_S \leq 10 \text{ k}\Omega$			6.0			7.5	mV
Input Offset Current				500			300	nA
Input Bias Current				1.5			0.8	μA
Large Signal Voltage Gain	$V_S = \pm15V$, $V_{OUT} = \pm10V$ $R_L \geq 2 \text{ k}\Omega$	25			25			V/mV
Output Voltage Swing	$V_S = \pm15V$, $R_L = 10 \text{ k}\Omega$ $R_L = 2 \text{ k}\Omega$	±12 ±10	±14 ±13		±12 ±10	±14 ±13		V V
Input Voltage Range	$V_S = \pm15V$	±12			±12			V
Common Mode Rejection Ratio	$R_S \leq 10 \text{ k}\Omega$	70	90		70	90		dB
Supply Voltage Rejection Ratio	$R_S \leq 10 \text{ k}\Omega$	77	96		77	96		dB

Note 1: The maximum junction temperature of the LM747 is $150°C$, while that of the LM747C is $100°C$. For operating at elevated temperatures, devices in the TO-5 package must be derated based on a thermal resistance of $150°C/W$, junction to ambient, or $45°C/W$, junction to case. For the flat package, the derating is based on a thermal resistance of $185°C/W$ when mounted on a 1/16-inch-thick epoxy glass board with ten, 0.03-inch-wide, 2-ounce copper conductors. The thermal resistance of the dual-in-line package is $100°C/W$, junction to ambient.

Note 2: For supply voltages less than ±15V, the absolute maximum input voltage is equal to the supply voltage.

Note 3: These specifications apply for $V_S = \pm15V$ and $-55°C \leq T_A \leq 125°C$, unless otherwise specified. With the LM747C, however, all specifications are limited to $0°C \leq T_A \leq 70°C$ $V_S = \pm15V$.

typical performance characteristics

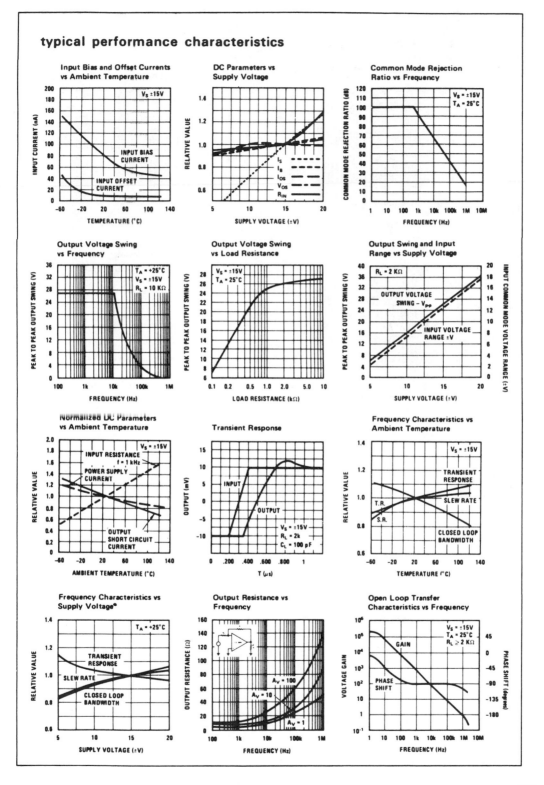

typical performance characteristics, continued

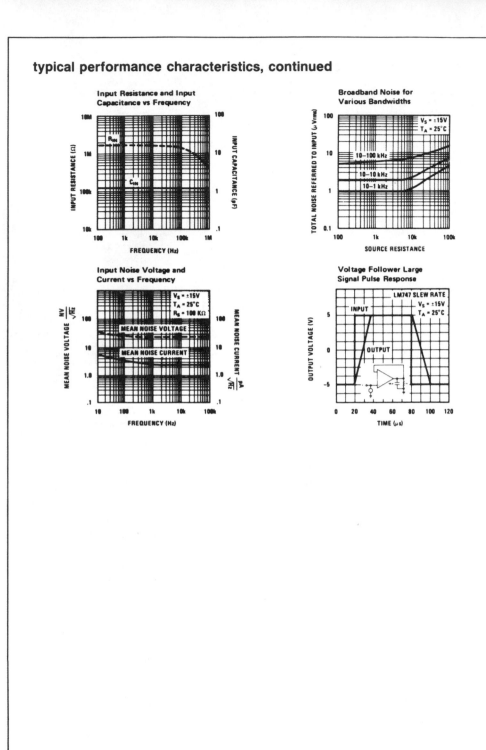

Input Resistance and Input Capacitance vs Frequency

Broadband Noise for Various Bandwidths

Input Noise Voltage and Current vs Frequency

Voltage Follower Large Signal Pulse Response

2. UNCOMPENSATED OPERATIONAL AMPLIFIER
(Reprinted by permission of National Semiconductor Corp.)

LM748/LM748C operational amplifier

general description

The LM748/LM748C is a general purpose operational amplifier built on a single silicon chip. The resulting close match and tight thermal coupling gives low offsets and temperature drift as well as fast recovery from thermal transients. In addition, the device features:

- Frequency compensation with a single 30 pF capacitor
- Operation from ±5V to ±20V
- Low current drain: 1.8 mA at ±20V
- Continuous short-circuit protection
- Operation as a comparator with differential inputs as high as ±30V

- No latch-up when common mode range is exceeded.
- Same pin configuration as the LM101.

The unity-gain compensation specified makes the circuit stable for all feedback configurations, even with capacitive loads. However, it is possible to optimize compensation for best high frequency performance at any gain. As a comparator, the output can be clamped at any desired level to make it compatible with logic circuits.

The LM748 is specified for operation over the –55°C to +125°C military temperature range. The LM748C is specified for operation over the 0°C to +70°C temperature range.

connection diagrams

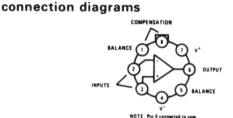

NOTE: Pin 4 connected to case

Order Number LM748H or LM748CH
See Package 11

TOP VIEW

Order Number LM748CN
See Package 20

typical applications

Inverting Amplifier with Balancing Circuit

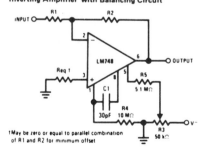

†May be zero or equal to parallel combination of R1 and R2 for minimum offset

Voltage Comparator for Driving DTL or TTL Integrated Circuits

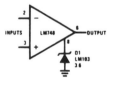

Low Drift Sample and Hold

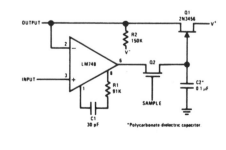

*Polycarbonate dielectric capacitor.

Voltage Comparator for Driving RTL Logic or High Current Driver

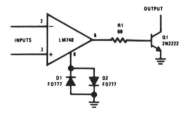

absolute maximum ratings

Supply Voltage	±22V
Power Dissipation (Note 1)	500 mW
Differential Input Voltage	±30V
Input Voltage (Note 2)	±15V
Output Short-Circuit Duration (Note 3)	Indefinite
Operating Temperature Range: LM748	-55°C to $+125^\circ$C
LM748C	0°C to $+70^\circ$C
Storage Temperature Range	-65°C to $+150^\circ$C
Lead Temperature (Soldering, 10 sec)	300°C

electrical characteristics (Note 4)

PARAMETER	CONDITIONS	MIN	TYP	MAX	UNITS
Input Offset Voltage	$T_A = 25^\circ$C, $R_S \leq 10$ kΩ		1.0	5.0	mV
Input Offset Current	$T_A = 25^\circ$C		40	200	nA
Input Bias Current	$T_A = 25^\circ$C		120	500	nA
Input Resistance	$T_A = 25^\circ$C	300	800		kΩ
Supply Current	$T_A = 25^\circ$C, $V_S = \pm15$V		1.8	2.8	mA
Large Signal Voltage Gain	$T_A = 25^\circ$C, $V_S = \pm15$V $V_{OUT} = \pm10$V, $R_L \geq 2$ kΩ	50	160		V/mV
Input Offset Voltage	$R_S \leq 10$ kΩ			6.0	mV
Average Temperature Coefficient of Input Offset Voltage	$R_S \leq 50\Omega$		3.0		μV/$^\circ$C
	$R_S \leq 10$ kΩ		6.0		μV/$^\circ$C
Input Offset Current	$T_A = 0^\circ$C to 70°C			300	nA
	$T_A = -55^\circ$C to 125°C			500	nA
Input Bias Current	$T_A = 0^\circ$C to 70°C			0.8	μA
	$T_A = -55^\circ$C to 125°C			1.5	μA
Supply Current	$T_A = +125^\circ$C, $V_S = \pm15$V		1.2	2.25	mA
	$T_A = -55^\circ$C to 125°C		1.9	3.3	mA
Large Signal Voltage Gain	$V_S = \pm15$V, $V_{OUT} = \pm10$V $R_L \geq 2$ KΩ	25			V/mV
Output Voltage Swing	$V_S = \pm15$V, $R_L = 10\Omega$	±12	±14		V
	$R_L = 2$ kΩ	±10	±13		V
Input Voltage Range	$V_S = \pm15$V	±12			V
Common Mode Rejection Ratio	$R_S \leq 10$ kΩ	70	90		dB
Supply Voltage Rejection Ratio	$R_S \leq 10$ kΩ	77	90		dB

Note 1: For operating at elevated temperatures the devices must be derated based on a maximum junction to case thermal resistance of 45°C per watt, or 150°C per watt junction to ambient. (See Curves).

Note 2: For supply voltages less than ±15V, the absolute maximum input voltage is equal to the supply voltage.

Note 3: Continuous short circuit is allowed for case temperatures to +125°C and ambient temperatures to +70°C.

Note 4: These specifications apply for ±5V $\leq V_S \leq$ +15V and -55°C $\leq T_A \leq$ 125°C, unless otherwise specified. With the LM748C, however, all temperature specifications are limited to 0°C $\leq T_A \leq$ 70°C.

3. VERY-HIGH-INPUT-IMPEDANCE OPERATIONAL AMPLIFIER

(Reprinted by permission of National Semiconductor Corp.)

LH0022/LH0022C* high performance FET op amp
LH0042/LH0042C low cost FET op amp
LH0052/LH0052C precision FET op amp

general description

The LH0022/LH0042/LH0052 are a family of FET input operational amplifiers with very closely matched input characteristics, very high input impedance, and ultra-low input currents with no compromise in noise, common mode rejection ratio, open loop gain, or slew rate. The internally laser nulled LH0052 offers 200 microvolts maximum offset and $5\,\mu V/^\circ C$ offset drift. Input offset current is less than 100 femtoamps at room temperature and 100 pA maximum at $125^\circ C$. The LH0022 and LH0042 are not internally nulled but offer comparable matching characteristics. All devices in the family are internally compensated and are free of latch-up and unusual oscillation problems. The devices may be offset nulled with a single 10k trimpot with neglible effect in offset drift or CMRR.

The LH0022, LH0042 and LH0052 are specified for operation over the $-55^\circ C$ to $+125^\circ C$ military temperature range. The LH0022C, LH0042C and LH0052C are specified for operation over the $-25^\circ C$ to $+85^\circ C$ temperature range.

features

- Low input offset current – 100 femtoamps max. (LH0052)
- Low input offset drift – $5\,\mu V/^\circ C$ max (LH0052)
- Low input offset voltage – 100 microvolts-typ.
- High open loop gain – 100 dB typ.
- Excellent slew rate – 3.0 V/µs typ.
- Internal 6 dB/octave frequency compensation
- Pin compatible with standard IC op amps (TO-5 package)

The LH0022/LH0042/LH0052 family of IC op amps are intended to fulfill a wide variety of applications for process control, medical instrumentation, and other systems requiring very low input currents and tightly matched input offsets. The LH0052 is particularly suited for long term high accuracy integrators and high accuracy sample and hold buffer amplifiers. The LH0022 and LH0042 provide low cost high performance for such applications as electrometer and photocell amplification, pico-ammeters, and high input impedance buffers.

Special electrical parameter selection and custom built circuits are available on special request.

For additional application information and information on other National operational amplifiers, see *Available Linear Applications Literature*.

schematic and connection diagrams

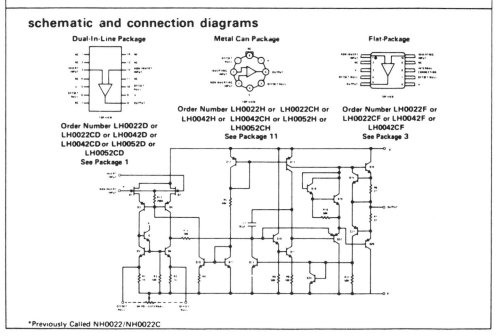

Dual-In-Line Package

Order Number LH0022D or LH0022CD or LH0042D or LH0042CD or LH0052D or LH0052CD
See Package 1

Metal Can Package

Order Number LH0022H or LH0042H or LH0042CH or LH0052H or LH0052CH
See Package 11

Flat-Package

Order Number LH0022F or LH0022CF or LH0042F or LH0042CF
See Package 3

*Previously Called NH0022/NH0022C

guaranteed performance characteristics (Note 4)

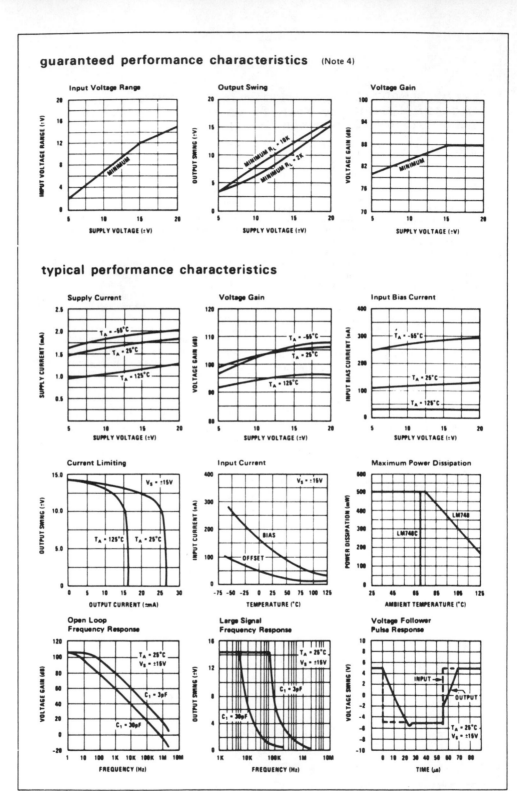

typical performance characteristics

absolute maximum ratings

Supply Voltage	±22V
Power Dissipation (see graph)	500 mW
Input Voltage (Note 1)	±15V
Differential Input Voltage (Note 2)	±30V
Voltage Between Offset Null and V⁻	±0.5V
Short Circuit Duration	Continuous
Operating Temperature Range	
LH0022, LH0042, LH0052	-55°C to $+125^\circ$C
LH0022C, LH0042C, LH0052C	-25°C to $+85^\circ$C
Storage Temperature Range	-65°C to $+150^\circ$C
Lead Temperature (Soldering, 10 sec)	300°C

dc electrical characteristics For LH0022/LH0022C (Note 3)

PARAMETER	CONDITIONS	LH0022 MIN	LH0022 TYP	LH0022 MAX	LH0022C MIN	LH0022C TYP	LH0022C MAX	UNITS
Input Offset Voltage	$R_S \leq 100$ kΩ; $T_A = 25^\circ$C		2.0	4.0		3.5	6.0	mV
	$R_S \leq 100$ kΩ			5.0			10.0	mV
Temperature Coefficient of Input Offset Voltage	$R_S \leq 100$ kΩ		5	10		5	15	μV/$^\circ$C
Offset Voltage Drift with Time			3			4		μV/week
Input Offset Current	$T_A = 25^\circ$C		0.2	2.0		1.0	5.0	pA
				200			200	pA
Temperature Coefficient of Input Offset Current		Doubles every 20°C			Doubles every 20°C			
Offset Current Drift with Time			0.1			0.1		pA/week
Input Bias Current	$T_A = 25^\circ$C		5	10		10	25	pA
				1.0			1.0	nA
Temperature Coefficient of Input Bias Current		Doubles every 20°C			Doubles every 20°C			
Differential Input Resistance			10^{12}			10^{12}		Ω
Common Mode Input Resistance			10^{12}			10^{12}		Ω
Input Capacitance			4.0			4.0		pF
Input Voltage Range	$V_S = \pm15$V	±12	±13.5		±12	±13.5		V
Common Mode Rejection Ratio	$R_S \leq 10$ kΩ, $V_{IN} = \pm10$V	80	90		70	90		dB
Supply Voltage Rejection Ratio	$R_S \leq 10$ kΩ, ±5V $\leq V_S \leq \pm15$V	80	90		70	90		dB
Large Signal Voltage Gain	$R_L = 2$ kΩ, $V_{OUT} = \pm10$V, $T_A = 25^\circ$C, $V_S = \pm15$V	100	200		75	160		V/mV
	$R_L = 2$ kΩ, $V_{OUT} = \pm10$V, $V_S = \pm15$V	50			50			V/mV
Output Voltage Swing	$R_L = 1$ kΩ, $T_A = 25^\circ$C, $V_S = \pm15$V	±10	±12.5		±10	±12		V
	$R_L = 2$ kΩ, $V_S = \pm15$V	±10			±10			V
Output Current Swing	$V_{OUT} = \pm10$V, $T_A = 25^\circ$C	±10	±15		±10	±15		mA
Output Resistance			75			75		Ω
Output Short Circuit Current			25			25		mA
Supply Current	$V_S = \pm15$V		2.0	2.5		2.4	2.8	mA
Power Consumption	$V_S = \pm15$V			75			85	mW

dc electrical characteristics for LH0042/LH0042C

($T_A = 25°C$, $V_S = ±15V$, unless otherwise specified)

PARAMETER	CONDITIONS	LH0042 MIN	LH0042 TYP	LH0042 MAX	LH0042C MIN	LH0042C TYP	LH0042C MAX	UNITS
Input Offset Voltage	$R_S ≤ 100\,k\Omega$, $±5V ≤ V_S ≤ 20V$		5.0	20		6.0	20	mV
Temperature Coefficient of Input Offset Voltage	$R_S ≤ 100\,k\Omega$		5	20		10	25	µV/°C
Offset Voltage Drift with Time			7			10		µV/week
Input Offset Current			1	5		2	10	pA
Temperature Coefficient of Input Offset Current			Doubles every 20°C			Doubles every 20°C		
Offset Current Drift with Time			0.1			0.1		pA/week
Input Bias Current			10	25		15	50	pA
Temperature Coefficient of Input Bias Current			Doubles every 20°C			Doubles every 20°C		
Differential Input Resistance			10^{12}			10^{12}		Ω
Common Mode Input Resistance			10^{12}			10^{12}		Ω
Input Capacitance			4.0			4.0		pF
Input Voltage Range		±12	±13.5		±12	±13.5		V
Common Mode Rejection Ratio	$R_S ≤ 10\,k\Omega$, $V_{IN} = ±10V$	70	86		70	80		dB
Supply Voltage Rejection Ratio	$R_S ≤ 10\,k\Omega$, $±5V ≤ V_S ≤ ±15V$	70	86		70	80		dB
Large Signal Voltage Gain	$R_L = 1\,k\Omega$, $V_{OUT} = ±10V$	50	150		25	100		V/mV
Output Voltage Swing	$R_L = 1\,k\Omega$	±10	±12.5		±10	±12		V
Output Current Swing	$V_{OUT} = ±10V$	±10	±15		±10	±15		mA
Output Resistance			75			75		Ω
Output Short Circuit Current			20			20		mA
Supply Current			2.5	3.5		2.8	4.0	mA
Power Consumption				105			120	mW

dc electrical characteristics For LH0052/LH0052C (Note 3)

PARAMETER	CONDITIONS	LH0052 MIN	LH0052 TYP	LH0052 MAX	LH0052C MIN	LH0052C TYP	LH0052C MAX	UNITS
Input Offset Voltage	$R_S ≤ 100\,k\Omega$, $V_S = ±15V$, $T_A = 25°C$		0.1	0.5		0.2	1.0	mV
	$R_S ≤ 100\,k\Omega$, $V_S = ±15V$			1.0			1.5	mV
Temperature Coefficient of Input Offset Voltage	$R_S ≤ 100\,k\Omega$		2	5		5	10	µV/°C
Offset Voltage Drift with Time			2			4		µV/week
Input Offset Current	$T_A = 25°C$		0.01	0.1		0.02	0.2	pA
				100			100	pA
Temperature Coefficient of Input Offset Current			Doubles every 20°C			Doubles every 20°C		
Offset Current Drift with Time			<0.1			<0.1		pA/week
Input Bias Current	$T_A = 25°C$		0.5	1.0		1.0	5.0	pA
				500			500	pA
Temperature Coefficient of Input Bias Current			Doubles every 20°C			Doubles every 20°C		
Differential Input Resistance			10^{12}			10^{12}		Ω
Common Mode Input Resistance			10^{12}			10^{12}		Ω
Input Capacitance			4.0			4.0		pF
Input Voltage Range	$V_S = ±15V$	±12	±13.5		±12	±13.5		V
Common Mode Rejection Ratio	$R_S ≤ 10\,k\Omega$, $V_{IN} = ±10V$	80	90		76	90		dB
Supply Voltage Rejection Ratio	$R_S ≤ 10\,k\Omega$, $±5V ≤ V_S ≤ ±15V$	80	90		76	90		dB
Large Signal Voltage Gain	$R_L = 2\,k\Omega$, $V_{OUT} = ±10V$, $V_S = ±15V$, $T_A = 25°C$	100	200		75	160		V/mV
	$R_L = 2\,k\Omega$, $V_{OUT} = ±10V$, $V_S = ±15V$	50			50			V/mV
Output Voltage Swing	$R_L = 1\,k\Omega$, $T_A = 25°C$, $V_S = ±15V$	±10	±12.5		±10	±12		V
	$R_L = 2\,k\Omega$, $V_S = ±15V$	±10			±10			V
Output Current Swing	$V_{OUT} = ±10V$, $T_A = 25°C$	±10	±15		±10	±15		mA
Output Resistance			75			75		Ω
Output Short Circuit Current			25			25		mA
Supply Current	$V_S = ±15V$		2.0	2.5		2.5	3.0	mA
Power Consumption	$V_S = ±15V$			75			90	mW

ac electrical characteristics For all amplifiers ($T_A = 25^\circ$C, $V_S = \pm15$V)

PARAMETER	CONDITIONS	LIMITS						UNITS
		LH0022/42/52			LH0022C/42C/52C			
		MIN	TYP	MAX	MIN	TYP	MAX	
Slew Rate	Voltage Follower	1.5	3.0		1.0	3.0		V/μs
Large Signal Bandwidth	Voltage Follower		40			40		kHz
Small Signal Bandwidth			1.0			1.0		MHz
Rise Time			0.3	1.5		0.3	1.5	μs
Overshoot			10	30		15	40	%
Settling Time (0.1 %)	$\Delta V_{IN} = 10$V		4.5			4.5		μs
Overload Recovery			4.0			4.0		μs
Input Noise Voltage	$R_S = 10$ kΩ, $f_o = 10$ Hz		150			150		nV/$\sqrt{Hz}$
Input Noise Voltage	$R_S = 10$ kΩ, $f_o = 100$ Hz		55			55		nV/$\sqrt{Hz}$
Input Noise Voltage	$R_S = 10$ kΩ, $f_o = 1$ kHz		35			35		nV/$\sqrt{Hz}$
Input Noise Voltage	$R_S = 10$ kΩ, $f_o = 10$ kHz		30			30		nV/$\sqrt{Hz}$
Input Noise Voltage	BW = 10 Hz to 10 kHz, $R_S = 10$ kΩ		12			12		μVrms
Input Noise Current	BW = 10 Hz to 10 kHz		<.1			<.1		pArms

171

typical performance characteristics

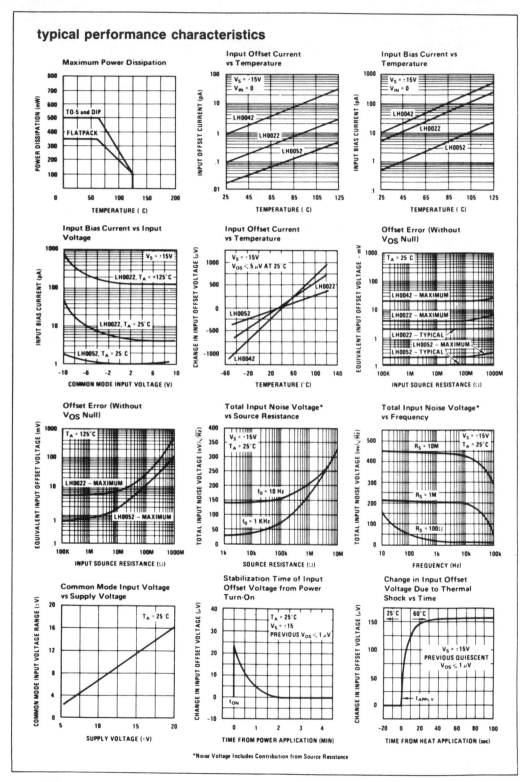

*Noise Voltage Includes Contribution from Source Resistance

typical performance characteristics, continued

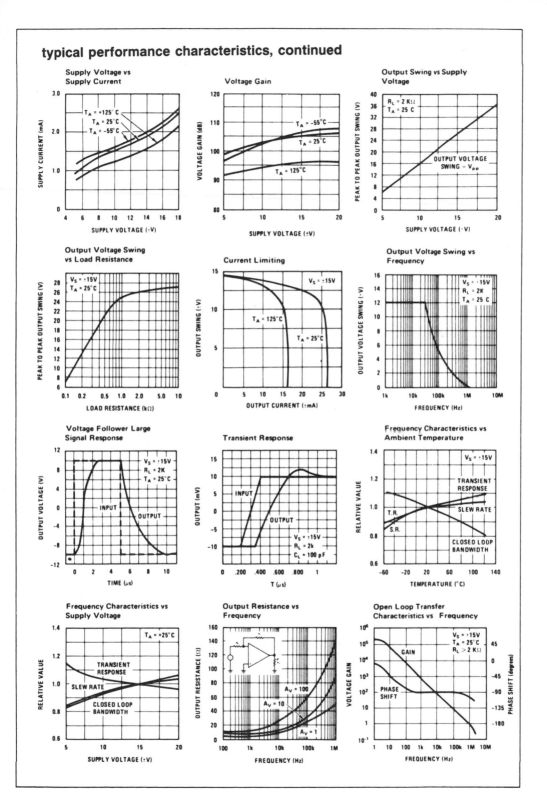

4. TRANSCONDUCTANCE OPERATIONAL AMPLIFIER

(Reprinted by permission of RCA Solid State Division, now Harris Semiconductor)

Operational Transconductance Amplifier

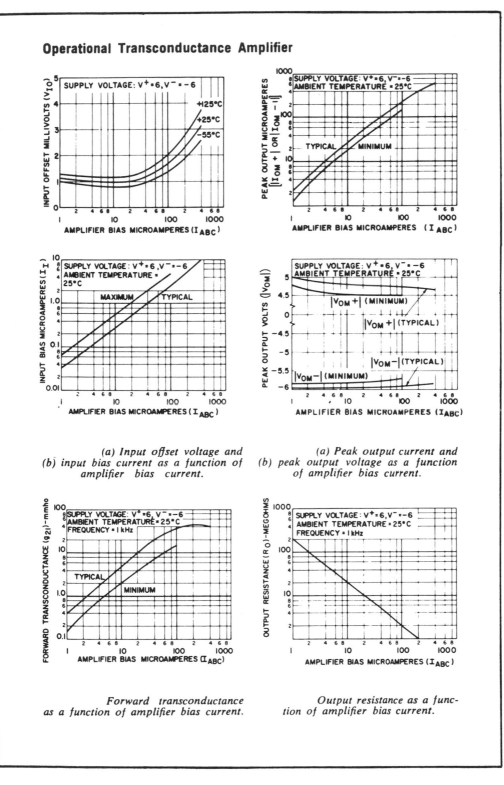

(a) Input offset voltage and (b) input bias current as a function of amplifier bias current.

(a) Peak output current and (b) peak output voltage as a function of amplifier bias current.

Forward transconductance as a function of amplifier bias current.

Output resistance as a function of amplifier bias current.

5. SINGLE-POWER-SUPPLY OPERATIONAL AMPLIFIER

(Reprinted by permission of National Semiconductor Corp.)

**National
Semiconductor
Corporation**

LM2900/LM3900, LM3301, LM3401 Quad Amplifiers

General Description

The LM2900 series consists of four independent, dual input, internally compensated amplifiers which were designed specifically to operate off of a single power supply voltage and to provide a large output voltage swing. These amplifiers make use of a current mirror to achieve the non-inverting input function. Application areas include: ac amplifiers, RC active filters, low frequency triangle, squarewave and pulse waveform generation circuits, tachometers and low speed, high voltage digital logic gates.

Features

- Wide single supply voltage   4 V_{DC} to 32 V_{DC}
  Range or dual supplies ± 2 V_{DC} to ± 16 V_{DC}
- Supply current drain independent of supply voltage
- Low input biasing current   30 nA
- High open-loop gain   70 dB
- Wide bandwidth   2.5 MHz (unity gain)
- Large output voltage swing $(V^+ - 1)$ Vp-p
- Internally frequency compensated for unity gain
- Output short-circuit protection

Schematic and Connection Diagrams

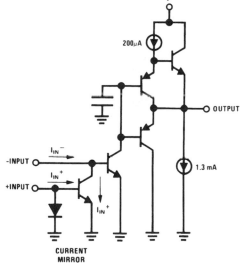

CURRENT
MIRROR

TL/H/7936-1

**Order Number LM3900M
See NS Package Number M14A**

Dual-In-Line and Flat Package

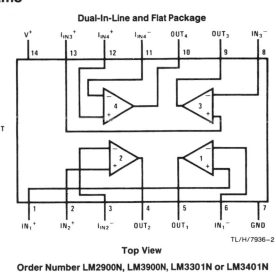

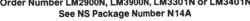

TL/H/7936-2

Top View

**Order Number LM2900N, LM3900N, LM3301N or LM3401N
See NS Package Number N14A**

Typical Applications (V$^+$ = 15 V$_{DC}$)

Inverting Amplifier

R1
100k
R2
1M
0.05 µF
LM3900
V$_O$
V$_{IN}$
2R2
2M
V$^+$

$$V_{ODC} = \frac{V^+}{2}$$

$$A_V \cong -\frac{R2}{R1}$$

TL/H/7936–3

Triangle/Square Generator

V$^+$
1M
0.001 µF
LM3900
V$_{O1}$
100K
1.8 msec
510K
LM3900
V$_{O2}$
V$^+$
1.2M
120K

TL/H/7936–4

Frequency-Doubling Tachometer

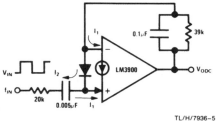

I$_1$
0.1µF
39k
V$_{IN}$
LM3900
V$_{ODC}$
I$_2$
f$_{IN}$
20k
0.005µF
I$_1$

TL/H/7936–5

Low V$_{IN}$ – V$_{OUT}$ Voltage Regulator

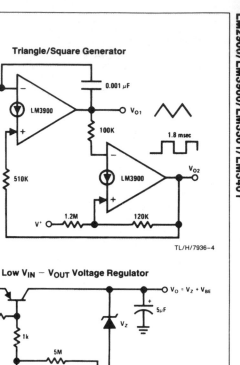

+V$_{IN}$
1k
1k
V$_O$ = V$_Z$ + V$_{BE}$
5µF
V$_Z$
5M
LM3900
+V$_{BE}$
510

TL/H/7936–6

Non-Inverting Amplifier

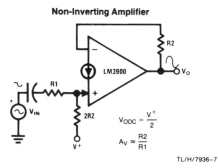

R2
LM3900
V$_O$
R1
V$_{IN}$
2R2
V$^+$

$$V_{ODC} = \frac{V^+}{2}$$

$$A_V \cong \frac{R2}{R1}$$

TL/H/7936–7

Negative Supply Biasing

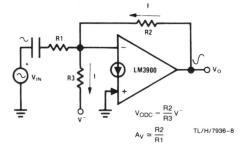

R2
R1
V$_{IN}$
LM3900
V$_O$
R3
V$^-$

$$V_{ODC} = \frac{R2}{R3}V^-$$

$$A_V \cong \frac{R2}{R1}$$

TL/H/7936–8

Absolute Maximum Ratings

If Military/Aerospace specified devices are required, contact the National Semiconductor Sales Office/Distributors for availability and specifications.

	LM2900/LM3900	LM3301	LM3401
Supply Voltage	32 V_{DC}	28 V_{DC}	18 V_{DC}
	$\pm 16 V_{DC}$	$\pm 14 V_{DC}$	$\pm 9 V_{DC}$
Power Dissipation (T_A = 25°C) (Note 1)	1220 mW		
Molded DIP	1080 mW	1080 mW	1080 mW
Input Currents, I_{IN}^+ or I_{IN}^-	20 mA_{DC}	20 mA_{DC}	20 mA_{DC}
Output Short-Circuit Duration—One Amplifier T_A = 25°C (See Application Hints)	Continuous	Continuous	Continuous
Operating Temperature Range		−40°C to +85°C	0°C to +75°C
LM2900	−40°C to +85°C		
LM3900	0°C to +70°C		
Storage Temperature Range	−65°C to +150°C	−65°C to +150°C	−65°C to +150°C
Lead Temperature (Soldering, 10 sec.)	260°C	260°C	260°C
Soldering Information			
Dual-In-Line Package			
Soldering (10 sec.)	260°C	260°C	260°C
Small Outline Package			
Vapor Phase (60 sec.)	215°C	215°C	215°C
Infrared (15 sec.)	220°C	220°C	220°C

See AN-450 "Surface Mounting Methods and Their Effect on Product Reliability" for other methods of soldering surface mount devices.

ESD rating to be determined.

Electrical Characteristics T_A = 25°C, V^+ = 15 V_{DC}, unless otherwise stated

Parameter		Conditions		LM2900			LM3900			LM3301			LM3401			Units
				Min	Typ	Max	Min	Typ	Max	Min	Typ	Max	Min	Typ	Max	
Open Loop	Voltage Gain	Over Temp.												0.8		V/mV
	Voltage Gain	ΔV_O = 10 V_{DC} Inverting Input		1.2	2.8		1.2	2.8		1.2	2.8		1.2	2.8		
	Input Resistance				1			1			1		0.1	1		MΩ
	Output Resistance				8			8			9			8		kΩ
Unity Gain Bandwidth		Inverting Input			2.5			2.5			2.5			2.5		MHz
Input Bias Current		Inverting Input, V^+ = 5 V_{DC} Inverting Input			30	200		30	200		30	300		30	300 500	nA
Slew Rate		Positive Output Swing Negative Output Swing			0.5 20			0.5 20			0.5 20			0.5 20		V/µs
Supply Current		R_L = ∞ On All Amplifiers			6.2	10		6.2	10		6.2	10		6.2	10	mA_{DC}
Output Voltage Swing	V_{OUT} High	R_L = 2k, V^+ = 15.0 V_{DC}	I_{IN}^- = 0, I_{IN}^+ = 0	13.5			13.5			13.5			13.5			
	V_{OUT} Low		I_{IN}^- = 10 µA, I_{IN}^+ = 0		0.09	0.2		0.09	0.2		0.09	0.2		0.09	0.2	V_{DC}
	V_{OUT} High	V^+ = Absolute Maximum Ratings R_L = ∞,	I_{IN}^- = 0, I_{IN}^+ = 0	29.5			29.5			26.0			16.0			
Output Current Capability	Source			6	18		6	10		5	18		5	10		
	Sink	(Note 2)		0.5	1.3		0.5	1.3		0.5	1.3		0.5	1.3		mA_{DC}
	I_{SINK}	V_{OL} = 1V, I_{IN}^- = 5 µA			5			5			5			5		

179

Electrical Characteristics (Note 6), $V^+ = 15\ V_{DC}$, unless otherwise stated (Continued)

Parameter	Conditions	LM2900			LM3900			LM3301			LM3401			Units
		Min	Typ	Max	Min	Typ	Max	Min	Typ	Max	Min	Typ	Max	
Power Supply Rejection	$T_A = 25°C$, f = 100 Hz		70			70			70			70		dB
Mirror Gain	@ 20 μA (Note 3)	0.90	1.0	1.1	0.90	1.0	1.1	0.90	1	1.10	0.90	1	1.10	$\mu A/\mu A$
	@ 200 μA (Note 3)	0.90	1.0	1.1	0.90	1.0	1.1	0.90	1	1.10	0.90	1	1.10	
ΔMirror Gain	@ 20 μA to 200 μA (Note 3)		2	5		2	5		2	5		2	5	%
Mirror Current	(Note 4)		10	500		10	500		10	500		10	500	μA_{DC}
Negative Input Current	$T_A = 25°C$ (Note 5)		1.0			1.0			1.0			1.0		mA_{DC}
Input Bias Current	Inverting Input		300			300								nA

Note 1: For operating at high temperatures, the device must be derated based on a 125°C maximum junction temperature and a thermal resistance of 92°C/W which applies for the device soldered in a printed circuit board, operating in a still air ambient.

Note 2: The output current sink capability can be increased for large signal conditions by overdriving the inverting input. This is shown in the section on Typical Characteristics.

Note 3: This spec indicates the current gain of the current mirror which is used as the non-inverting input.

Note 4: Input V_{BE} match between the non-inverting and the inverting inputs occurs for a mirror current (non-inverting input current) of approximately 10 μA. This is therefore a typical design center for many of the application circuits.

Note 5: Clamp transistors are included on the IC to prevent the input voltages from swinging below ground more than approximately $-0.3\ V_{DC}$. The negative input currents which may result from large signal overdrive with capacitance input coupling need to be externally limited to values of approximately 1 mA. Negative input currents in excess of 4 mA will cause the output voltage to drop to a low voltage. This maximum current applies to any one of the input terminals. If more than one of the input terminals are simultaneously driven negative smaller maximum currents are allowed. Common-mode current biasing can be used to prevent negative input voltages; see for example, the "Differentiator Circuit" in the applications section.

Note 6: These specs apply for $-40°C \leq T_A \leq +85°C$, unless otherwise stated.

Application Hints

When driving either input from a low-impedance source, a limiting resistor should be placed in series with the input lead to limit the peak input current. Currents as large as 20 mA will not damage the device, but the current mirror on the non-inverting input will saturate and cause a loss of mirror gain at mA current levels—especially at high operating temperatures.

Precautions should be taken to insure that the power supply for the integrated circuit never becomes reversed in polarity or that the unit is not inadvertently installed backwards in a test socket as an unlimited current surge through the resulting forward diode within the IC could cause fusing of the internal conductors and result in a destroyed unit.

Output short circuits either to ground or to the positive power supply should be of short time duration. Units can be destroyed, not as a result of the short circuit current causing metal fusing, but rather due to the large increase in IC chip dissipation which will cause eventual failure due to excessive junction temperatures. For example, when operating from a well-regulated +5 V_{DC} power supply at $T_A = 25°C$ with a 100 kΩ shunt-feedback resistor (from the output to the inverting input) a short directly to the power supply will not cause catastrophic failure but the current magnitude will be approximately 50 mA and the junction temperature will be above T_J max. Larger feedback resistors will reduce the current, 11 MΩ provides approximately 30 mA, an open circuit provides 1.3 mA, and a direct connection from the output to the non-inverting input will result in catastrophic failure when the output is shorted to V^+ as this then places the base-emitter junction of the input transistor directly across the power supply. Short-circuits to ground will have magnitudes of approximately 30 mA and will not cause catastrophic failure at $T_A = 25°C$.

Unintentional signal coupling from the output to the non-inverting input can cause oscillations. This is likely only in breadboard hook-ups with long component leads and can be prevented by a more careful lead dress or by locating the non-inverting input biasing resistor close to the IC. A quick check of this condition is to bypass the non-inverting input to ground with a capacitor. High impedance biasing resistors used in the non-inverting input circuit make this input lead highly susceptible to unintentional AC signal pickup.

Operation of this amplifier can be best understood by noticing that input currents are differenced at the inverting-input terminal and this difference current then flows through the external feedback resistor to produce the output voltage. Common-mode current biasing is generally useful to allow operating with signal levels near ground or even negative as this maintains the inputs biased at $+V_{BE}$. Internal clamp transistors (see note 5) catch-negative input voltages at approximately $-0.3\ V_{DC}$ but the magnitude of current flow has to be limited by the external input network. For operation at high temperature, this limit should be approximately 100 μA.

This new "Norton" current-differencing amplifier can be used in most of the applications of a standard IC op amp. Performance as a DC amplifier using only a single supply is not as precise as a standard IC op amp operating with split supplies but is adequate in many less critical applications. New functions are made possible with this amplifier which are useful in single power supply systems. For example, biasing can be designed separately from the AC gain as was shown in the "inverting amplifier," the "difference integrator" allows controlling the charging and the discharging of the integrating capacitor with positive voltages, and the "frequency doubling tachometer" provides a simple circuit which reduces the ripple voltage on a tachometer output DC voltage.

Typical Performance Characteristics

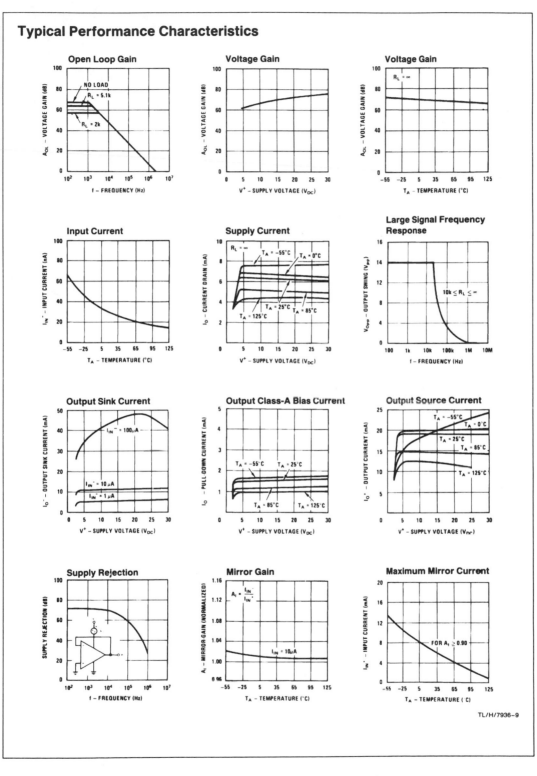

TL/H/7936-9

6. LOW-VOLTAGE-POWER-SUPPLY OPERATIONAL AMPLIFIER
(Reprinted by permission of Intersil.)

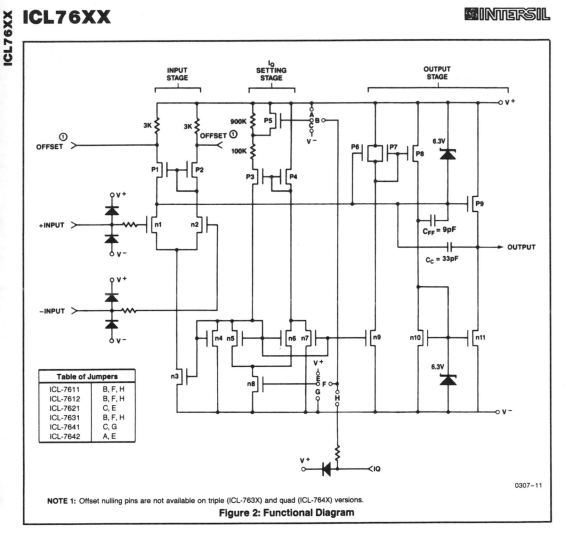

Table of Jumpers

ICL-7611	B, F, H
ICL-7612	B, F, H
ICL-7621	C, E
ICL-7631	B, F, H
ICL-7641	C, G
ICL-7642	A, E

0307-11

NOTE 1: Offset nulling pins are not available on triple (ICL-763X) and quad (ICL-764X) versions.

Figure 2: Functional Diagram

NOTE: All typical values have been characterized but are not tested.

Device	Description	Pin Assignments

ICL7621XCPA
ICL7621XCTV
ICL7621XMTV
	Dual op amps with internal compensation; I_Q fixed at 100µA
	Pin compatible with
	  Texas Inst. TL082
	  Motorola MC1458
	  Raytheon RC4558
		TO-99 (TOP VIEW)
		(outline dwg TV)
		
		* PIN DIP (TOP VIEW)
		(outline dwg PA)
		
		0307-6 0307-7
		*Pin 8 connected to case.

ICL7631XCPE	Triple op amps with internal compensation. Adjustable I_Q Same pin configuration as ICL8023.	16 PIN DIP (TOP VIEW) (outline dwgs JE, PE) 0307-9 Note: pins 5 and 15 are internally connected.

ICL7641XCPD
ICL7642XCPD
	Quad op amps with internal compensation.
	I_Q fixed at 1mA (ICL7641)
	I_Q fixed at 10µA (ICL7642)
	Pin compatible with
	  Texas Instr. TL084
	  National LM324
	  Harris HA4741
		14 PIN DIP (TOP VIEW)
		(outline dwg JD, PD)
		
		0307-10

Figure 1: Pin Configurations (Cont.)

NOTE: All typical values have been characterized but are not tested.

ICL76XX
ICL76XX Series Low Power CMOS Operational Amplifiers

GENERAL DESCRIPTION

The ICL761X/762X/763X/764X series is a family of monolithic CMOS operational amplifiers. These devices provide the designer with high performance operation at low supply voltages and selectable quiescent currents, and are an ideal design tool when ultra low input current and low power dissipation are desired.

The basic amplifier will operate at supply voltages ranging from $\pm 1V$ to $\pm 8V$, and may be operated from a single Lithium cell.

A unique quiescent current programming pin allows setting of standby current to 1mA, 100μA, or 10μA, with no external components. This results in power consumption as low as 20μW. Output swings range to within a few millivolts of the supply voltages.

Of particular significance is the extremely low (1pA) input current, input noise current of $.01pA/\sqrt{Hz}$, and $10^{12}\Omega$ input impedance. These features optimize performance in very high source impedance applications.

The inputs are internally protected and require no special handling procedures. Outputs are fully protected against short circuits to ground or to either supply.

AC performance is excellent, with a slew rate of 1.6V/μs, and unity gain bandwidth of 1MHz at $I_Q = 1mA$.

Because of the low power dissipation, operating temperatures and drift are quite low. Applications utilizing these features may include stable instruments, extended life designs, or high density packages.

FEATURES

- Wide Operating Voltage Range $\pm 1V$ to $\pm 8V$
- High Input Impedance — $10^{12}\Omega$
- Programmable Power Consumption — Low As 20μW
- Input Current Lower Than BIFETs — Typ 1pA
- Available As Singles, Duals, Triples, and Quads
- Output Voltage Swings to Within Millivolts Of V$^-$ and V$^+$
- Low Power Replacement for Many Standard Op Amps
- Compensated and Uncompensated Versions
- Input Common Mode Voltage Range Greater Than Supply Rails (ICL7612)

APPLICATIONS

- Portable Instruments
- Telephone Headsets
- Hearing Aid/Microphone Amplifiers
- Meter Amplifiers
- Medical Instruments
- High Impedance Buffers

SELECTION GUIDE

DEVICE NOMENCLATURE

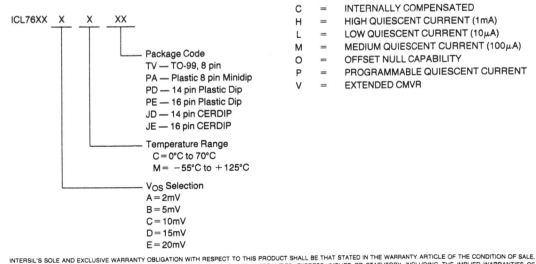

ICL76XX X X XX

- Package Code
  - TV — TO-99, 8 pin
  - PA — Plastic 8 pin Minidip
  - PD — 14 pin Plastic Dip
  - PE — 16 pin Plastic Dip
  - JD — 14 pin CERDIP
  - JE — 16 pin CERDIP
- Temperature Range
  - C = 0°C to 70°C
  - M = −55°C to +125°C
- V_{OS} Selection
  - A = 2mV
  - B = 5mV
  - C = 10mV
  - D = 15mV
  - E = 20mV

SPECIAL FEATURE CODES

C	=	INTERNALLY COMPENSATED
H	=	HIGH QUIESCENT CURRENT (1mA)
L	=	LOW QUIESCENT CURRENT (10μA)
M	=	MEDIUM QUIESCENT CURRENT (100μA)
O	=	OFFSET NULL CAPABILITY
P	=	PROGRAMMABLE QUIESCENT CURRENT
V	=	EXTENDED CMVR

302060−002

NOTE: All typical values have been characterized but are not tested.

ABSOLUTE MAXIMUM RATINGS

Total Supply Voltage V^+ to V^- 18V
Input Voltage .... V^- -0.3 to V^+ $+0.3$V
Differential Input Voltage[1] . $\pm[(V^+ +0.3) - (V^- -0.3)]$V
Duration of Output Short Circuit[2] .... Unlimited

Continuous Power Dissipation

	@25°C	Above 25°C derate as below:
TO-99	250mW	2mW/°C
8 Lead Minidip	250mW	2mW/°C
14 Lead Plastic	375mW	3mW/°C
14 Lead Cerdip	500mW	4mW/°C
16 Lead Plastic	375mW	3mW/°C
16 Lead Cerdip	500mW	4mW/°C

Storage Temperature Range .... $-65°C$ to $+150°C$
Operating Temperature Range
ICL76XXM .... $-55°C$ to $+125°C$
ICL76XXC .... 0°C to $+70°C$
Lead Temperature (Soldering, 10sec) .... 300°C

NOTE: *Stresses above those listed under "Absolute Maximum Ratings" may cause permanent damage to the device. These are stress ratings only and functional operation of the device at these or any other conditions above those indicated in the operational sections of the specifications is not implied. Exposure to absolute maximum rating conditions for extended periods may affect device reliability.*

NOTE 1. Long term offset voltage stability will be degraded if large input differential voltages are applied for long periods of time.
   2. The outputs may be shorted to ground or to either supply. for $V_{SUPP} \leq 10V$. Care must be taken to insure that the dissipation rating is not exceeded.

ELECTRICAL CHARACTERISTICS (7611/12 and 7621 ONLY)

($V_{SUPPLY} = \pm5.0V$, $T_A = 25°C$, unless otherwise specified.)

Symbol	Parameter	Test Conditions	76XXA Min	76XXA Typ	76XXA Max	76XXB Min	76XXB Typ	76XXB Max	76XXD Min	76XXD Typ	76XXD Max	Units
V_{OS}	Input Offset Voltage	$R_S \leq 100k\Omega$, $T_A = 25°C$ $T_{MIN} \leq T_A \leq T_{MAX}$			2 3			5 7			15 20	mV
$\Delta V_{OS}/\Delta T$	Temperature Coefficient of V_{OS}	$R_S \leq 100k\Omega$		10			15			25		$\mu V/°C$
I_{OS}	Input Offset Current	$T_A = 25°C$ $\Delta T_A = C_{(2)}$ $\Delta T_A = M_{(2)}$		0.5	30 300 800		0.5	30 300 800		0.5	30 300 800	pA
I_{BIAS}	Input Bias Current	$T_A = 25°C$ $\Delta T_A = C$ $\Delta T_A = M$		1.0	50 400 4000		1.0	50 400 4000		1.0	50 400 4000	pA
V_{CMR}	Common Mode Voltage Range (Except ICL7612)	$I_Q = 10\mu A^{(1)}$ $I_Q = 100\mu A$ $I_Q = 1mA^{(1)}$	±4.4 ±4.2 ±3.7			±4.4 ±4.2 ±3.7			±4.4 ±4.2 ±3.7			V
V_{CMR}	Extended Common Mode Voltage Range (ICL7612 Only)	$I_Q = 10\mu A$	±5.3			±5.3			±5.3			V
		$I_Q = 100\mu A$	$+5.3$ -5.1			$+5.3$ -5.1			$+5.3$ -5.1			
		$I_Q = 1mA$	$+5.3$ -4.5			$+5.3$ -4.5			$+5.3$ -4.5			
V_{OUT}	Output Voltage Swing	(1) $I_Q = 10\mu A$, $R_L = 1M\Omega$ $T_A = 25°C$ $\Delta T_A = C$ $\Delta T_A = M$	±4.9 ±4.8 ±4.7			±4.9 ±4.8 ±4.7			±4.9 ±4.8 ±4.7			V
		$I_Q = 100\mu A$, $R_L = 100k\Omega$ $T_A = 25°C$ $\Delta T_A = C$ $\Delta T_A = M$	±4.9 ±4.8 ±4.5			±4.9 ±4.8 ±4.5			±4.9 ±4.8 ±4.5			V
		(1) $I_Q = 1mA$, $R_L = 10k\Omega$ $T_A = 25°C$ $\Delta T_A = C$ $\Delta T_A = M$	±4.5 ±4.3 ±4.0			±4.5 ±4.3 ±4.0			±4.5 ±4.3 ±4.0			

NOTE: All typical values have been characterized but are not tested.

ICL76XX

ELECTRICAL CHARACTERISTICS (7611/12 and 7621 ONLY) (Continued)

($V_{SUPPLY} = \pm 5.0V$, $T_A = 25°C$, unless otherwise specified.)

Symbol	Parameter	Test Conditions	76XXA Min	76XXA Typ	76XXA Max	76XXB Min	76XXB Typ	76XXB Max	76XXD Min	76XXD Typ	76XXD Max	Units
A_{VOL}	Large Signal Voltage Gain	$V_O = \pm 4.0V$, $R_L = 1M\Omega$ $I_Q = 10\mu A^{(1)}$, $T_A = 25°C$ $\Delta T_A = C$ $\Delta T_A = M$	86 80 74	104		80 75 68	104		80 75 68	104		dB
		$V_O = \pm 4.0V$, $R_L = 100k\Omega$ $I_Q = 100\mu A$, $T_A = 25°C$ $\Delta T_A = C$ $\Delta T_A = M$	86 80 74	102		80 75 68	102		80 75 68	102		
		$V_O = \pm 4.0V$, $R_L = 10k\Omega$ $I_Q = 1mA^{(1)}$, $T_A = 25°C$ $\Delta T_A = C$ $\Delta T_A = M$	80 76 72	83		76 72 68	83		76 72 68	83		
GBW	Unity Gain Bandwidth	$I_Q = 10\mu A^{(1)}$ $I_Q = 100\mu A$ $I_Q = 1mA^{(1)}$		0.044 0.48 1.4			0.044 0.48 1.4			0.044 0.48 1.4		MHz
R_{IN}	Input Resistance			10^{12}			10^{12}			10^{12}		Ω
CMRR	Common Mode Rejection Ratio	$R_S \le 100k\Omega$, $I_Q = 10\mu A^{(1)}$ $R_S \le 100k\Omega$, $I_Q = 100\mu A$ $R_S \le 100k\Omega$, $I_Q = 1mA^{(1)}$	76 76 66	96 91 87		70 70 60	96 91 87		70 70 60	96 91 87		dB
PSRR	Power Supply Rejection Ratio	$R_S \le 100k\Omega$, $I_Q = 10\mu A^{(1)}$ $R_S \le 100k\Omega$, $I_Q = 100\mu A$ $R_S \le 100k\Omega$, $I_Q = 1mA^{(1)}$	80 80 70	94 86 77		80 80 70	94 86 77		80 80 70	94 86 77		dB
e_n	Input Referred Noise Voltage	$R_S = 100\Omega$, $f = 1kHz$		100			100			100		$nV/\sqrt{Hz}$
i_n	Input Referred Noise Current	$R_S = 100\Omega$, $f = 1kHz$		0.01			0.01			0.01		$pA/\sqrt{Hz}$
I_{SUPPLY}	Supply Current (Per Amplifier)	No Signal, No Load I_Q SET = $+5V^{(1)}$ I_Q SET = 0V I_Q SET = $-5V^{(1)}$		0.01 0.1 1.0	0.02 0.25 2.5		0.01 0.1 1.0	0.02 0.25 2.5		0.01 0.1 1.0	0.02 0.25 2.5	mA
V_{O1}/V_{O2}	Channel Separation	$A_{VOL} = 100$		120			120			120		dB
SR	Slew Rate[3]	$A_{VOL} = 1$, $C_L = 100pF$ $V_{IN} = 8Vp-p$ $I_Q = 10\mu A^{(1)}$, $R_L = 1M\Omega$ $I_Q = 100\mu A$, $R_L = 100k\Omega$ $I_Q = 1mA^{(1)}$, $R_L = 10k\Omega$		0.016 0.16 1.6			0.016 0.16 1.6			0.016 0.16 1.6		$V/\mu s$
t_r	Rise Time[3]	$V_{IN} = 50mV$, $C_L = 100pF$ $I_Q = 10\mu A^{(1)}$, $R_L = 1M\Omega$ $I_Q = 100\mu A$, $R_L = 100k\Omega$ $I_Q = 1mA^{(1)}$, $R_L = 10k\Omega$		20 2 0.9			20 2 0.9			20 2 0.9		μs
	Overshoot Factor[3]	$V_{IN} = 50mV$, $C_L = 100pF$ $I_Q = 10\mu A^1$, $R_L = 1M\Omega$ $I_Q = 100\mu A$, $R_L = 100k\Omega$ $I_Q = 1mA^1$, $R_L = 10k\Omega$		5 10 40			5 10 40			5 10 40		%

NOTES: 1. ICL7611, 7612 only.

2. C = Commercial Temperature Range: 0°C to +70°C
M = Military Temperature Range: −55°C to +125°C

NOTE: All typical values have been characterized but are not tested.

ELECTRICAL CHARACTERISTICS (7611/12 AND 7621 ONLY)

($V_{SUPPLY} = \pm 1.0V$, $I_Q = 10\mu A$, $T_A = 25°C$, unless otherwise specified.)

Symbol	Parameter	Test Conditions	76XXA			76XXB			Units
			Min	Typ	Max	Min	Typ	Max	
V_{OS}	Input Offset Voltage	$R_S \leq 100k\Omega$, $T_A = 25°C$ $T_{MIN} \leq T_A \leq T_{MAX}$			2 3			5 7	mV
$\Delta V_{OS}/\Delta T$	Temperature Coefficient of V_{OS}	$R_S \leq 100k\Omega$		10			15		$\mu V/°C$
I_{OS}	Input Offset Current	$T_A = 25°C$ $\Delta T_A = C$		0.5	30 300		0.5	30 300	pA
I_{BIAS}	Input Bias Current	$T_A = 25°C$ $\Delta T_A = C$		1.0	50 500		1.0	50 500	pA
V_{CMR}	Common Mode Voltage Range (Except ICL7612)		± 0.6			± 0.6			V
V_{CMR}	Extended Common Mode Voltage Range (ICL7612 Only)		$+0.6$ to -1.1			$+0.6$ to -1.1			V
V_{OUT}	Output Voltage Swing	$R_L = 1M\Omega$, $T_A = 25°C$ $\Delta T_A = C$		± 0.98 ± 0.96			± 0.98 ± 0.96		V
A_{VOL}	Large Signal Voltage Gain	$V_O = \pm 0.1V$, $R_L = 1M\Omega$ $T_A = 25°C$ $\Delta T_A = C$		90 80			90 80		dB
GBW	Unity Gain Bandwidth			0.044					MHz
R_{IN}	Input Resistance			10^{12}			10^{12}		
CMRR	Common Mode Rejection Ratio	$R_S \leq 100k\Omega$		80			80		
PSRR	Power Supply Rejection Ratio	$R_S \leq 100k\Omega$		80			80		dB
e_n	Input Referred Noise Voltage	$R_S = 100\Omega$, $f = 1kHz$		100			100		$nV/\sqrt{Hz}$
i_n	Input Referred Noise Current	$R_S = 100\Omega$, $f = 1kHz$		0.01			0.01		$pA/\sqrt{Hz}$
I_{SUPPLY}	Supply Current (Per Amplifier)	No Signal, No Load		6	15		6	15	μA
SR	Slew Rate	$A_{VOL} = 1$, $C_L = 100pF$ $V_{IN} = 0.2Vp\text{-}p$ $R_L = 1M\Omega$		0.016			0.016		$V/\mu s$
t_r	Rise Time	$V_{IN} = 50mV$, $C_L = 100pF$ $R_L = 1M\Omega$		20			20		μs
	Overshoot Factor	$V_{IN} = 50mV$, $C_L = 100pF$ $R_L = 1M\Omega$		5			5		%

NOTE: C = Commercial Temperature Range (0°C to +70°C)   M = Military Temperature Range (−55°C to +125°C).

NOTE: All typical values have been characterized but are not tested.

ELECTRICAL CHARACTERISTICS (7631, 7641/42 ONLY)

($V_{SUPPLY} = \pm 5.0V$, $T_A = 25°C$, unless otherwise specified.)

Symbol	Parameter	Test Conditions	76XXC (6)			76XXE (6)			Units
			Min	Typ	Max	Min	Typ	Max	
V_{OS}	Input Offset Voltage	$R_S \le 100k\Omega$, $T_A = 25°C$			10			20	mV
		$T_{MIN} \le T_A \le T_{MAX}$			15			25	
$\Delta V_{OS}/\Delta T$	Temperature Coefficient of V_{OS}	$R_S \le 100k\Omega$ (Note 5)		20			30		
I_{OS}	Input Offset Current	$T_A = 25°C$		0.5	30		0.5	30	
		$\Delta T_A = C$			300			300	pA
		$\Delta T_A = M$			800			800	
I_{BIAS}	Input Bias Current	$T_A = 25°C$		1.0	50		1.0	50	
		$\Delta T_A = C$			500			500	pA
		$\Delta T_A = M$			4000			4000	
V_{CMR}	Common Mode Voltage Range	$I_Q = 10\mu A^{(1)}$	± 4.4			± 4.4			
		$I_Q = 100\mu A^{(3)}$	± 4.2			± 4.2			V
		$I_Q = 1mA^{(2)}$	± 3.7			± 3.7			
V_{OUT}	Output Voltage Swing	(1) $I_Q = 10\mu A$, $R_L = 1M\Omega$							
		$T_A = 25°C$	± 4.9			± 4.9			
		$\Delta T_A = C$	± 4.8			± 4.8			
		$\Delta T_A = M$	± 4.7			± 4.7			
		$I_Q = 100\mu A$, $R_L = 100k\Omega$							
		(3) $T_A = 25°C$	± 4.9			± 4.9			V
		$\Delta T_A = C$	± 4.8			± 4.8			
		$\Delta T_A = M$	± 4.5			± 4.5			
		(2) $I_Q = 1mA$, $R_L = 10k\Omega$							
		$T_A = 25°C$	± 4.5			± 4.5			
		$\Delta T_A = C$	± 4.3			± 4.3			
		$\Delta T_A = M$	± 4.0			± 4.0			
A_{VOL}	Large Signal Voltage Gain	$V_O = \pm 4.0V$, $R_L = 1M\Omega^{(1)}$							
		$I_Q = 10\mu A^{(1)}$, $T_A = 25°C$	80	104		80	104		
		$\Delta T_A = C$	75			75			
		$\Delta T_A = M$	68			68			
		$V_O = \pm 4.0V$, $R_L = 100k\Omega^{(3)}$							
		$I_Q = 100\mu A$, $T_A = 25°C$	80	102		80	102		
		$\Delta T_A = C$	75			75			
		$\Delta T_A = M$	68			68			dB
		$V_Q = \pm 4.0V$, $R_L = 10k\Omega^{(2)}$							
		$I_Q = 1mA^{(1)}$, $T_A = 25°C$	80	98		80	98		
		$\Delta T_A = C$	75			75			
		$\Delta T_A = M$	68			68			
GBW	Unity Gain Bandwidth	$I_Q = 10\mu A^{(1)}$		0.044			0.044		MHz
		$I_Q = 100\mu A^{(3)}$		0.48			0.48		
		$I_Q = 1mA^{(2)}$		1.4			1.4		
R_{IN}	Input Resistance			10^{12}			10^{12}		Ω
CMRR	Common Mode Rejection Ratio	$R_S \le 100k\Omega$, $I_Q = 10\mu A^{(1)}$	70	96		70	96		
		$R_S \le 100k\Omega$, $I_Q = 100\mu A$	70	91		70	91		dB
		$R_S \le 100k\Omega$, $I_Q = 1mA^{(2)}$	60	87		60	87		

NOTE: All typical values have been characterized but are not tested.

ICL76XX

TYPICAL PERFORMANCE CHARACTERISTICS

SUPPLY CURRENT PER AMPLIFIER AS A FUNCTION OF SUPPLY VOLTAGE

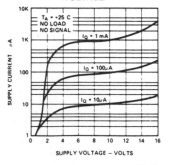

0307-12

SUPPLY CURRENT PER AMPLIFIER AS A FUNCTION OF FREE-AIR TEMPERATURE

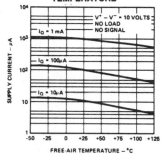

0307-13

INPUT BIAS CURRENT AS A FUNCTION OF TEMPERATURE

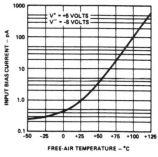

0307-14

LARGE SIGNAL DIFFERENTIAL VOLTAGE GAIN AS A FUNCTION OF FREE-AIR TEMPERATURE

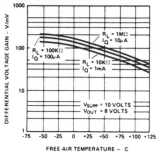

0307-15

LARGE SIGNAL DIFFERENTIAL VOLTAGE GAIN AND PHASE SHIFT AS A FUNCTION OF FREQUENCY

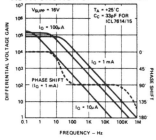

0307-16

COMMON MODE REJECTION RATIO AS A FUNCTION OF FREE-AIR TEMPERATURE

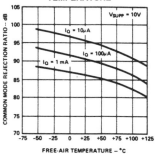

0307-17

POWER SUPPLY REJECTION RATIO AS A FUNCTION OF FREE-AIR TEMPERATURE

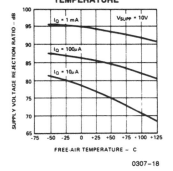

0307-18

EQUIVALENT INPUT NOISE VOLTAGE AS A FUNCTION OF FREQUENCY

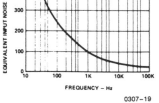

0307-19

PEAK-TO-PEAK OUTPUT VOLTAGE AS A FUNCTION OF FREQUENCY

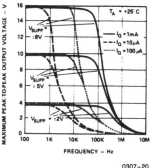

0307-20

NOTE: All typical values have been characterized but are not tested.

190

ⓢINTERSIL

TYPICAL PERFORMANCE CHARACTERISTICS

MAXIMUM PEAK-TO-PEAK OUTPUT VOLTAGE AS A FUNCTION OF FREQUENCY

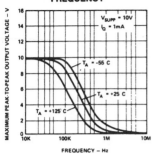

0307-21

MAXIMUM PEAK-TO-PEAK OUTPUT VOLTAGE AS A FUNCTION OF SUPPLY VOLTAGE

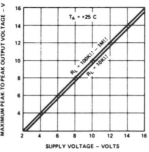

0307-22

MAXIMUM PEAK-TO-PEAK VOLTAGE AS A FUNCTION OF FREE-AIR TEMPERATURE

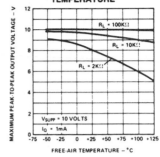

0307-23

MAXIMUM OUTPUT SOURCE CURRENT AS A FUNCTION OF SUPPLY VOLTAGE

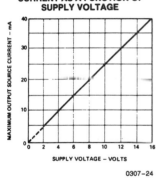

0307-24

MAXIMUM OUTPUT SINK CURRENT AS A FUNCTION OF SUPPLY VOLTAGE

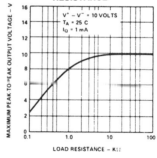

0307-25

MAXIMUM PEAK-TO-PEAK OUTPUT VOLTAGE AS A FUNCTION OF LOAD RESISTANCE

0307-26

VOLTAGE FOLLOWER LARGE SIGNAL PULSE RESPONSE

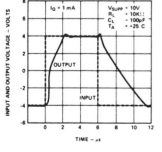

0307-27

VOLTAGE FOLLOWER LARGE SIGNAL PULSE RESPONSE

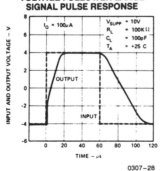

0307-28

VOLTAGE FOLLOWER LARGE SIGNAL PULSE RESPONSE

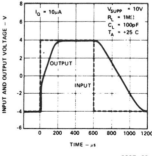

0307-29

NOTE: All typical values have been characterized but are not tested.

7. VERY-LOW-VOLTAGE-POWER-SUPPLY OPERATIONAL AMPLIFIER

(Reprinted by permission of National Semiconductor Corp.)

![National Semiconductor Corporation]

LM4250/LM4250C Programmable Operational Amplifier

General Description

The LM4250 and LM4250C are extremely versatile programmable monolithic operational amplifiers. A single external master bias current setting resistor programs the input bias current, input offset current, quiescent power consumption, slew rate, input noise, and the gain-bandwidth product. The device is a truly general purpose operational amplifier.

The LM4250C is identical to the LM4250 except that the LM4250C has its performance guaranteed over a 0°C to +70°C temperature range instead of the −55°C to +125°C temperature range of the LM4250.

Features

- ±1V to ±18V power supply operation
- 3 nA input offset current
- Standby power consumption as low as 500 nW
- No frequency compensation required
- Programmable electrical characteristics
- Offset voltage nulling capability
- Can be powered by two flashlight batteries
- Short circuit protection

Typical Applications

X5 Difference Amplifier

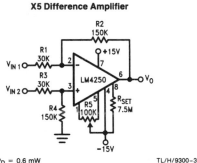

Quiescent P_D = 0.6 mW

TL/H/9300–3

500 Nano-Watt X10 Amplifier

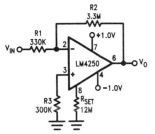

TL/H/9300–4

Quiescent P_P = 500 mW

Connection Diagrams

Metal Can Package

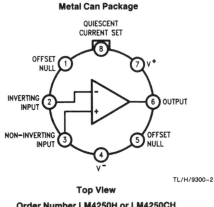

Top View

TL/H/9300–2

Order Number LM4250H or LM4250CH
See NS Package Number H08C

Dual-In-Line Package

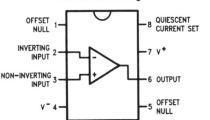

Top View

TL/H/9300–5

Order Number LM4250J, LM4250CJ,
LM4250CN or LM4250M
See NS Package Number J08A, M08A or N08E

Absolute Maximum Ratings

If Military/Aerospace specified devices are required, contact the National Semiconductor Sales Office/Distributors for availability and specifications.
(Note 2)

	LM4250	LM4250C
Supply Voltage	±18V	±18V
Operating Temp. Range	$-55°C \leq T_A \leq +125°C$	$0°C \leq T_A \leq +70°C$
Differential Input Voltage	±30V	±30V
Input Voltage (Note 1)	±15V	±15V
I_{SET} Current	150 nA	150 nA
Output Short Circuit Duration	Indefinite	Indefinite
T_{JMAX}		
H-Package	150°C	100°C
N-Package		100°C
J-Package	150°C	100°C
M-Package		100°C
Power Dissipation at $T_A = 25°C$		
H-Package (Still Air)	500 mW	300 mW
(400 LF/Min Air Flow)	1200 mW	1200 mW
N-Package		500 mW
J-Package	1000 mW	600 mW
M-Package		350 mW
Thermal Resistance (Typical) θ_{JA}		
H-Package (Still Air)	225°C/W	225°C/W
(400 LF/Min Air Flow)	90°C/W	90°C/W
N-Package		130°C/W
J-Package	108°C/W	108°C/W
M-Package		190°C/W
(Typical) θ_{JC}		
H-Package (Still Air)	25°C/W	25°C/W
(400 LF/Min Air Flow)	10°C/W	10°C/W
Storage Temperature Range	$-65°C$ to $+150°C$	$-65°C$ to $+150°C$

Soldering Information
  Dual-In-Line Package
    Soldering (10 seconds)   260°C
  Small Outline Package
    Vapor Phase (60 seconds)   215°C
    Infrared (15 seconds)   220°C

See AN-450 "Surface Mounting Methods and Their Effect on Product Reliability" for other methods of soldering surface mount devices.

ESD rating to be determined.

Note 1: For supply voltages less than ±15V, the absolute maximum input voltage is equal to the supply voltage.

Note 2: Refer to RETS4250X for military specifications.

Resistor Biasing

Set Current Setting Resistor to V⁻

	I_{SET}				
V_S	0.1 μA	0.5 μA	1.0 μA	5 μA	10 μA
±1.5V	25.6 MΩ	5.04 MΩ	2.5 MΩ	492 kΩ	244 kΩ
±3.0V	55.6 MΩ	11.0 MΩ	5.5 MΩ	1.09 MΩ	544 kΩ
±6.0V	116 MΩ	23.0 MΩ	11.5 MΩ	2.29 MΩ	1.14 MΩ
±9.0V	176 MΩ	35.0 MΩ	17.5 MΩ	3.49 MΩ	1.74 MΩ
±12.0V	236 MΩ	47.0 MΩ	23.5 MΩ	4.69 MΩ	2.34 MΩ
±15.0V	296 MΩ	59.0 MΩ	29.5 MΩ	5.89 MΩ	2.94 MΩ

Electrical Characteristics LM4250 ($-55°C \leq T_A \leq +125°C$ unless otherwise specified.) $T_A = T_J$

| Parameter | Conditions | $V_S = \pm 1.5V$ | | | |
| | | $I_{SET} = 1\,\mu A$ | | $I_{SET} = 10\,\mu A$ | |
		Min	Max	Min	Max
V_{OS}	$R_S \leq 100\,k\Omega$, $T_A = 25°C$		3 mV		5 mV
I_{OS}	$T_A = 25°C$		3 nA		10 nA
I_{bias}	$T_A = 25°C$		7.5 nA		50 nA
Large Signal Voltage Gain	$R_L = 100\,k\Omega$, $T_A = 25°C$	40k			
	$V_O = \pm 0.6V$, $R_L = 10\,k\Omega$			50k	
Supply Current	$T_A = 25°C$		7.5 μA		80 μA
Power Consumption	$T_A = 25°C$		23 μW		240 μW
V_{OS}	$R_S \leq 100\,k\Omega$		4 mV		6 mV
I_{OS}	$T_A = +125°C$		5 nA		10 nA
	$T_A = -55°C$		3 nA		10 nA
I_{bias}			7.5 nA		50 nA
Input Voltage Range		$\pm 0.6V$		$\pm 0.6V$	
Large Signal Voltage Gain	$V_O = \pm 0.5V$, $R_L = 100\,k\Omega$	30k			
	$R_L = 10\,k\Omega$			30k	
Output Voltage Swing	$R_L = 100\,k\Omega$	$\pm 0.6V$			
	$R_L = 10\,k\Omega$			$\pm 0.6V$	
Common Mode Rejection Ratio	$R_S \leq 10\,k\Omega$	70 dB		70 dB	
Supply Voltage Rejection Ratio	$R_S \leq 10\,k\Omega$	76 dB		76 dB	
Supply Current			8 μA		90 μA
Power Consumption			24 μW		270 μW

| Parameter | Conditions | $V_S = \pm 15V$ | | | |
| | | $I_{SET} = 1\,\mu A$ | | $I_{SET} = 10\,\mu A$ | |
		Min	Max	Min	Max
V_{OS}	$R_S \leq 100\,k\Omega$, $T_A = 25°C$		3 mV		5 mV
I_{OS}	$T_A = 25°C$		3 nA		10 nA
I_{bias}	$T_A = 25°C$		7.5 nA		50 nA
Large Signal Voltage Gain	$R_L = 100\,k\Omega$, $T_A = 25°C$	100k			
	$V_O = \pm 10V$, $R_L = 10\,k\Omega$			100k	
Supply Current	$T_A = 25°C$		10 μA		90 μA
Power Consumption	$T_A = 25°C$		300 μW		2.7 mW
V_{OS}	$R_S \leq 100\,k\Omega$		4 mV		6 mV
I_{OS}	$T_A = +125°C$		25 nA		25 nA
	$T_A = -55°C$		3 nA		10 nA
I_{bias}			7.5 nA		50 nA
Input Voltage Range		$\pm 13.5V$		$\pm 13.5V$	
Large Signal Voltage Gain	$V_O = \pm 10V$, $R_L = 100\,k\Omega$	50k			
	$R_L = 10\,k\Omega$			50k	
Output Voltage Swing	$R_L = 100\,k\Omega$	$\pm 12V$			
	$R_L = 10\,k\Omega$			$\pm 12V$	
Common Mode Rejection Ratio	$R_S \leq 10\,k\Omega$	70 dB		70 dB	
Supply Voltage Rejection Ratio	$R_S \leq 10\,k\Omega$	76 dB		76 dB	
Supply Current			11 μA		100 μA
Power Consumption			330 μW		3 mW

Electrical Characteristics LM4250C ($-55°C \leq T_A \leq +125°C$ unless otherwise specified.) $T_A = T_J$

Parameter	Conditions	$V_S = \pm 1.5V$			
		$I_{SET} = 1\,\mu A$		$I_{SET} = 10\,\mu A$	
		Min	Max	Min	Max
V_{OS}	$R_S \leq 100\,k\Omega$, $T_A = 25°C$		5 mV		6 mV
I_{OS}	$T_A = 25°C$		6 nA		20 nA
I_{bias}	$T_A = 25°C$		10 nA		75 nA
Large Signal Voltage Gain	$R_L = 100\,k\Omega$, $T_A = 25°C$ $V_O = \pm 0.6V$, $R_L = 10\,k\Omega$	25k		25k	
Supply Current	$T_A = 25°C$		8 μA		90 μA
Power Consumption	$T_A = 25°C$		24 μW		270 μW
V_{OS}	$R_S \leq 10\,k\Omega$		6.5 mV		7.5 mV
I_{OS}			8 nA		25 nA
I_{bias}			10 nA		80 nA
Input Voltage Range		$\pm 0.6V$		$\pm 0.6V$	
Large Signal Voltage Gain	$V_O = \pm 0.5V$, $R_L = 100\,k\Omega$ $R_L = 10\,k\Omega$	25k		25k	
Output Voltage Swing	$R_L = 100\,k\Omega$ $R_L = 10\,k\Omega$	$\pm 0.6V$		$\pm 0.6V$	
Common Mode Rejection Ratio	$R_S \leq 10\,k\Omega$	70 dB		70 dB	
Supply Voltage Rejection Ratio	$R_S \leq 10\,k\Omega$	74 dB		74 dB	
Supply Current			8 μA		90 μA
Power Consumption			24 μW		270 μW

Parameter	Conditions	$V_S = \pm 15V$			
		$I_{SET} = 1\,\mu A$		$I_{SET} = 10\,\mu A$	
		Min	Max	Min	Max
V_{OS}	$R_S \leq 100\,k\Omega$, $T_A = 25°C$		5 mV		6 mV
I_{OS}	$T_A = 25°C$		6 nA		20 nA
I_{bias}	$T_A = 25°C$		10 nA		75 nA
Large Signal Voltage Gain	$R_L = 100\,k\Omega$, $T_A = 25°C$ $V_O = \pm 10V$, $R_L = 10\,k\Omega$	60k		60k	
Supply Current	$T_A = 25°C$		11 μA		100 μA
Power Consumption	$T_A = 25°C$		330 μW		3 mW
V_{OS}	$R_S \leq 100\,k\Omega$		6.5 mV		7.5 mV
I_{OS}			8 nA		25 nA
I_{bias}			10 nA		80 nA
Input Voltage Range		$\pm 13.5V$		$\pm 13.5V$	
Large Signal Voltage Gain	$V_O = \pm 10V$, $R_L = 100\,k\Omega$ $R_L = 10\,k\Omega$	50k		50k	
Output Voltage Swing	$R_L = 100\,k\Omega$ $R_L = 10\,k\Omega$	$\pm 12V$		$\pm 12V$	
Common Mode Rejection Ratio	$R_S \leq 10\,k\Omega$	70 dB		70 dB	
Supply Voltage Rejection Ratio	$R_S \leq 10\,k\Omega$	74 dB		74 dB	
Supply Current			11 μA		100 μA
Power Consumption			330 μW		3 mW

Typical Performance Characteristics

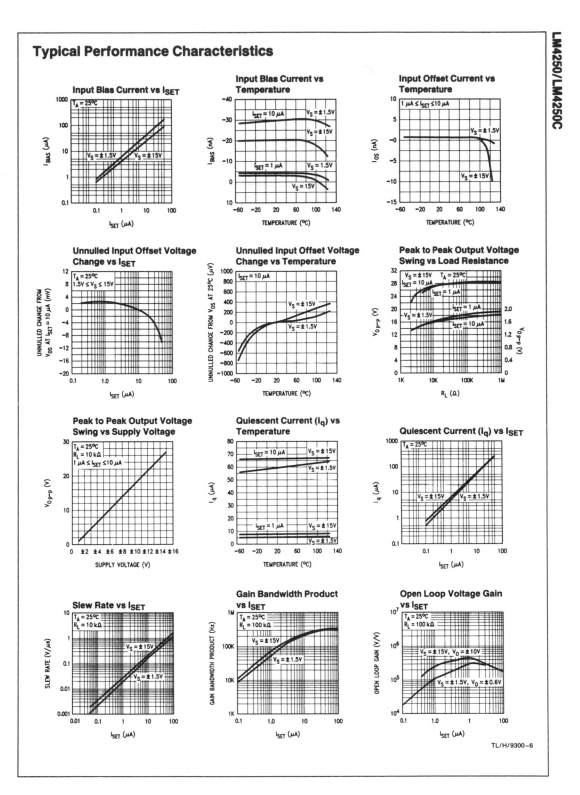

Typical Performance Characteristics (Continued)

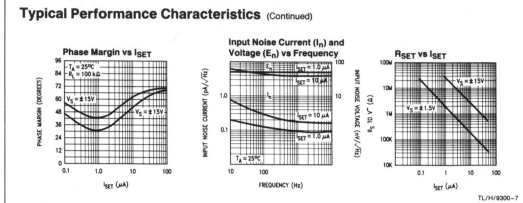

TL/H/9300–7

Typical Applications (Continued)

Floating Input Meter Amplifier
100 nA Full Scale

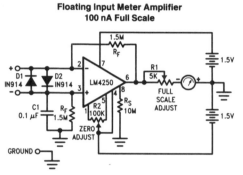

Quiescent P_D = 1.8 μW

TL/H/9300–8

*Meter movement (0–100 μA, 2 kΩ) marked
for 0–100 nA full scale.

X100 Instrumentation Amplifier 10 μW

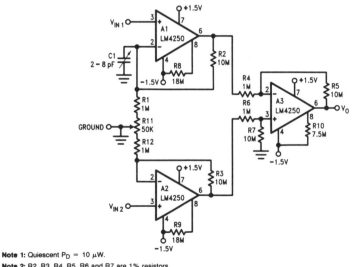

Note 1: Quiescent P_D = 10 μW.

Note 2: R2, R3, R4, R5, R6 and R7 are 1% resistors.

Note 3: R11 and C1 are for DC and AC common mode rejection adjustments.

TL/H/9300–9

8. LOW-NOISE OPERATIONAL AMPLIFIER

(Reprinted by permission of RCA Solid State Division, now Harris Semiconductor)

CA6078, CA6741

MAXIMUM RATINGS, *Absolute-Maximum Values at $T_A = 25°C$*

	CA6741T	CA6078AT
DC Supply Voltage (between V+ and V− terminals)	44 V	36 V
Differential-Mode Input Voltage	±30 V	±6 V
Common-Mode DC Input Voltage▲	±15 V	V+ to V−
Device Dissipation:		
Up to 75°C (CA6741T), Up to 125° (CA6078AT)	500 mW	250 mW
Above 75°C	Derate linearly 5 mW/°C	−
Temperature Range:		
Operating	−55 to +125 °C	−55 to +125 °C
Storage	−65 to +150 °C	−65 to +150 °C
Output Short-Circuit Duration●	No limitation	No limitation
Lead Temperature (During soldering):		
At distance 1/16 ±1/32 inch (1.59 ±0.79 mm)		
from case for 10 seconds max.	300 °C	300 °C

▲If Supply Voltage is less than ±15 volts, the Absolute Maximum Input Voltage is equal to the Supply Voltage.

●Short circuit may be applied to ground or to either supply.

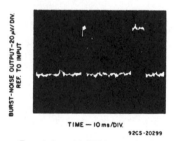

a. *Typ. device with high-burst-noise characteristic.*

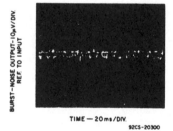

b. *Typ. device controlled for burst noise.*

Fig.1—*Typ. waveforms of type with high burst noise and type controlled for burst noise.*

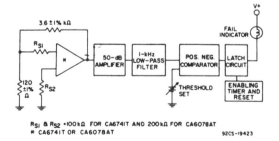

R_{S1} & R_{S2} =100kΩ FOR CA6741T AND 200kΩ FOR CA6078AT
* CA6741T OR CA6078AT   92CS-19423

Fig.2—*Block diagram of burst-noise "popcorn" test equipment.*

Operational Amplifiers

CA6078AT — Micropower Type
CA6741T — General-Purpose Type

For Applications where Low Noise (Burst + 1/f) is a Prime Requirement

Virtually free from "popcorn" (burst) noise:
device rejected if any noise burst exceeds 20 µV (peak),
referred to input over a 30-second time period.

8-LEAD TO-5
with Dual-In-Line
Formed Leads

8-LEAD
TO-5

H-1787   H-1528

RCA-CA6078AT and CA6741T* are low-noise linear IC operational amplifiers that are virtually free of "popcorn" (burst) noise.

These low-noise versions of the CA3078AT and CA3741T are a result of improved processing developments and rigid burst-noise inspection criteria. A highly selective test circuit (See Fig. 2) assures that each type meets the rigid low-noise standards shown in the data section. This low-burst-noise property also assures excellent performance throughout the 1/f noise spectrum.

In addition the CA6078AT and CA6741T offer the same features incorporated in the CA3078AT and CA3741T respectively, including output short-circuit protection, latch-free operation, wide common-mode and differential-mode signal ranges, and low-offset nulling capability.

For detailed data, characteristics curves, schematic diagram, dimensional outline, and test circuits, refer to the Operational Amplifier Data Bulletins File No. 531 and 535. In addition, for details of considerations in burst-noise measurements, refer to Application Note, ICAN-6732, "Measurement of Burst ("Popcorn") Noise in Linear IC's".

The CA6078AT and CA6741T utilize the hermetically sealed 8-lead TO-5 type package. The CA6078AT and the CA6741T can also be supplied on request with dual-in-line formed leads. These types are identified as the CA6078AS and CA6741S. This formed-lead configuration conforms to that of the 8-lead dual-in-line (Mini—Dip) package. For terminal arrangements, see page 4.

*Formerly Dev. No. TA5807X and TA6029 respectively.

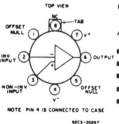

CA6741T

NOTE PIN 4 IS CONNECTED TO CASE

92CS-20297

Applications:

- **Low-noise AC amplifier**
- **Narrow-band or band-pass filter**
- **Integrator or differentiator**
- **DC amplifier**
- **Summing amplifier**

Features:

- **Internal phase compensation**
- **Input bias current: 500 nA max.**
- **Input offset current: 200 nA max.**
- **Open-loop voltage gain: 50,000 (94 dB) min.**
- **Input offset voltage: 5 mV max.**

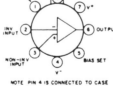

CA6078AT

NOTE PIN 4 IS CONNECTED TO CASE

92CS-20298

Applications:

- **Portable electronics**
- **Medical electronics**
- **DC amplifier**
- **Narrow-band or band-pass filter**
- **Integrator or differentiator**
- **Instrumentation**
- **Telemetry**
- **Summing amplifier**

Features:

- **Open-loop voltage gain: 40,000 (92 dB) min.**
- **Input offset voltage: 3.5 mV max.**
- **Operates with low total supply voltage: 1.5 V min. (± 0.75 V)**
- **Low quiescent operating current: adjustable for application optimization**
- **Input bias current: adjustable to below 1 nA**

201

ELECTRICAL CHARACTERISTICS – CA6078AT, *For Equipment Design.*

CHARACTERISTICS	SYMBOLS	TEST CONDITIONS Supply Volts: V^+ = 6, V^- = –6 T_A = 25°C, I_Q = 20 μA	LIMITS			UNITS
			MIN.	TYP.	MAX.	
Noise Characteristic						
"Popcorn" (Burst) Noise		Bandwidth = 1 kHz R_{SI} = R_{S2} = 200 kΩ	Device is rejected if the total noise voltage (burst + 1/$_f$), referred to input, exceeds 20 μV peak, during a 30-sec. test period.			
Principal Characteristics (For detailed Electrical Characteristics refer to CA3078AT Data Bulletin, File No. 535.)						
Input Offset Voltage	V_{IO}	$R_S \leq$ 10 kΩ	–	0.7	3.5	mV
Input Offset Current	I_{IO}		–	0.5	2.5	nA
Input Bias Current	I_{IB}		–	7	12	nA
Open-Loop Differential Voltage Gain	A_{OL}	$R_L \geq$ 10 kΩ V_O = ±4V	40,000 92	100,000 100	– –	dB
Common-Mode Input Voltage Range	V_{ICR}	$V^+ = V^-$ = 15 V	±14	–	–	V
Common-Mode Rejection Ratio	CMRR	$R_S \leq$ 10 kΩ	80	115	–	dB
Output Voltage Swing	$V_{O(P\text{-}P)}$	$R_L \geq$ 10 Ω	±13.7	±14.1	–	V
		$R_L \geq$ 2 kΩ	–	±14	–	
Supply Current	I_Q		–	20	25	μA

ELECTRICAL CHARACTERISTICS – CA6741T, *For Equipment Design.*

CHARACTERISTICS	SYMBOLS	TEST CONDITIONS Supply Volts; V^+ = 15, V^- = –15 T_A = 25°C	LIMITS			UNITS
			MIN.	TYP.	MAX.	
Noise Characteristic						
"Popcorn" (Burst) Noise		Bandwidth = 1 kHz R_{S1} = R_{S2} = 100 kΩ	Device is rejected if the total noise voltage (burst + 1/$_f$), referred to input, exceeds 20 μV peak, during a 30-sec. test period.			
Principal Characteristics (For detailed Electrical Characteristics refer to CA3741T Data Bulletin, File No. 531.)						
Input Offset Voltage	V_{IO}	$R_S \leq$ 10 kΩ	–	1	5	mV
Input Offset Current	I_{IO}		–	20	200	nA
Input Bias Current	I_{IB}		–	80	500	nA
Open-Loop Differential Voltage Gain	A_{OL}	$R_L \geq$ 2 kΩ V_O = ±10 V	50,000 94	200,000 106	– –	dB
Common-Mode Input Voltage Range	V_{ICR}		±12	±13	–	V
Common-Mode Rejection Ratio	CMRR	$R_S \leq$ 10 kΩ	70	90	–	dB
Output Voltage Swing	$V_{O(P\text{-}P)}$	$R_L \geq$ 10 kΩ	±12	±14	–	V
		$R_L \geq$ 2 kΩ	±10	±13	–	
Supply Current	I_Q		–	1.7	2.8	mA

CA6078, CA6741

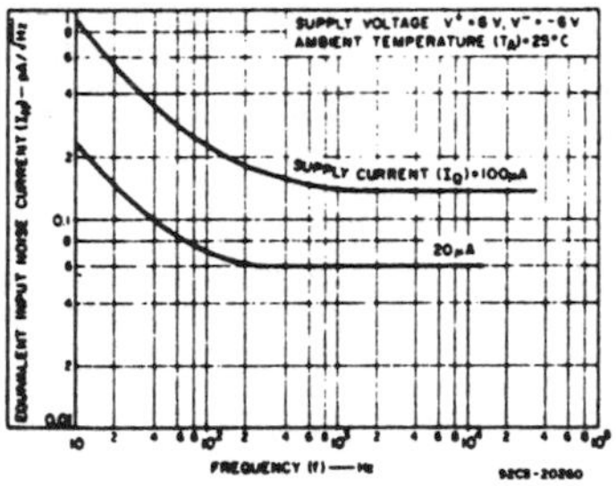

Fig.3–I_N vs. Frequency for CA6078AT.

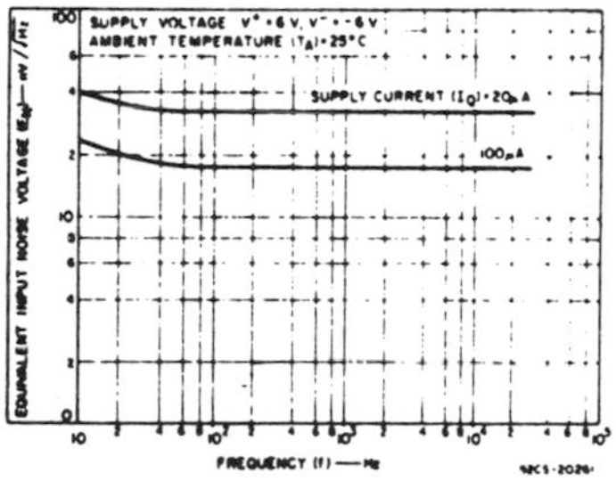

Fig.4–E_N vs. Frequency for CA6078AT.

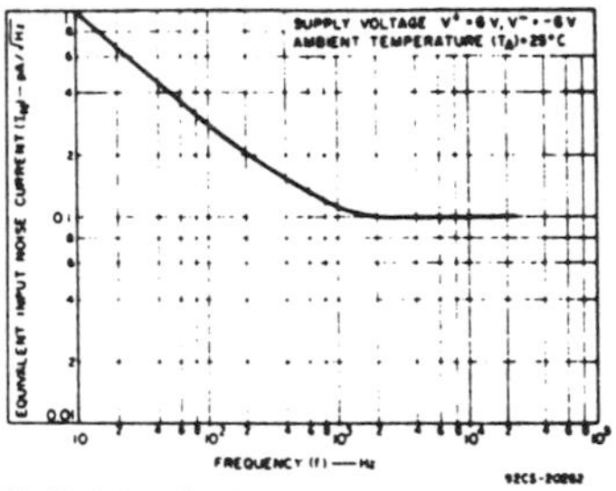

Fig.5–I_N vs. Frequency for CA6741T.

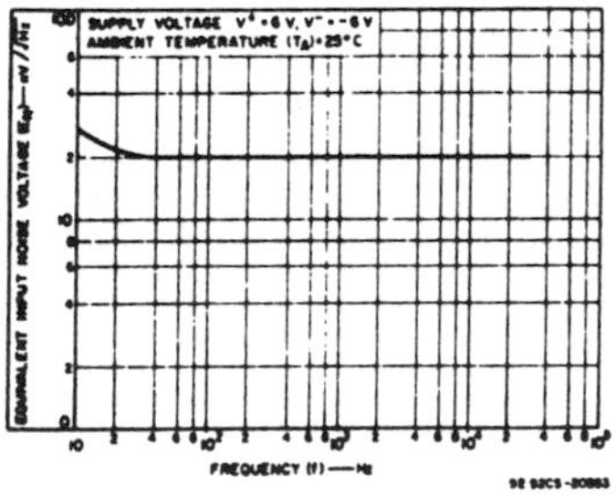

Fig.6–E_N vs. Frequency for CA6741T.

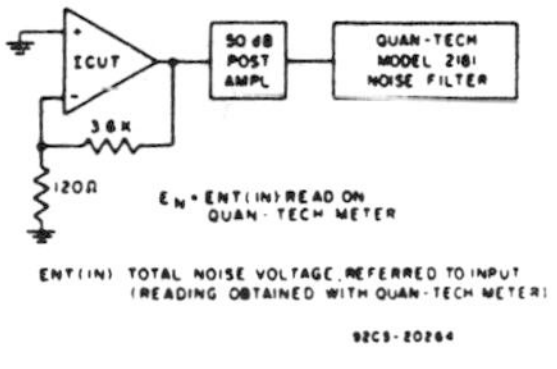

Fig.7–Test block diagram for E_N.

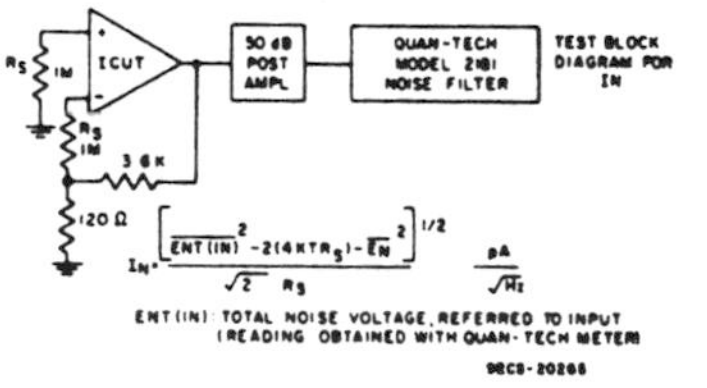

Fig.8–Test block diagram for I_N.

9. WIDEBAND OPERATIONAL AMPLIFIER

(Reprinted by permission of RCA Solid State Division, now Harris Semiconductor)

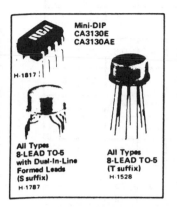

Mini-DIP
CA3130E
CA3130AE

H-1817

All Types
8-LEAD TO-5
with Dual-In-Line
Formed Leads
(S suffix)
H-1787

All Types
8-LEAD TO-5
(T suffix)
H-1528

BiMOS
Operational Amplifiers

With MOS/FET Input/ COS/MOS Output

FEATURES:

- *MOS/FET input stage provides:*
  very high Z = 1.5 TΩ $(1.5 \times 10^{12}Ω)$ typ.
  very low II = 5 pA typ. at 15-V operation
  2 pA typ. at 5-V operation
- *Common-mode input-voltage range includes negative supply rail; input terminals can be swung 0.5 V below negative supply rail*
- *COS/MOS output stage permits signal swing to either (or both) supply rails*

} *Ideal for single-supply applications*

RCA-CA3130T, CA3130E, CA3130S, CA-3130AT, CA 3130AS, CA3130AE, CA3130BT, and CA3130BS are integrated-circuit operational amplifiers that combine the advantages of both COS/MOS and bipolar transistors on a monolithic chip.

Gate-protected p-channel MOS/FET (PMOS) transistors are used in the input circuit to provide very-high-input impedance, very-low-input current, and exceptional speed performance. The use of PMOS field-effect transistors in the input stage results in common-mode input-voltage capability down to 0.5 volt below the negative-supply terminal, an important attribute in single-supply applications.

A complementary-symmetry MOS (COS/MOS) transistor-pair, capable of swinging the output voltage to within 10 millivolts of either supply-voltage terminal (at very high values of load impedance), is employed as the output circuit.

The CA3130 Series circuits operate at supply voltages ranging from 5 to 16 volts, or ±2.5 to ±8 volts when using split supplies. They can be phase compensated with a single external capacitor, and have terminals for adjustment of offset voltage for applications requiring offset-null capability. Terminal provisions are also made to permit strobing of the output stage.

The CA3130 Series is supplied in standard 8-lead TO-5 style packages (T suffix), 8-lead dual-in-line formed lead TO-5 style "DIL-CAN" packages (S suffix). The CA3130 is available in chip form (H suffix). The CA3130 and CA3130A are also available in the Mini-DIP 8-lead dual-in-line plastic

- *Low V_{IO}: 2 mV max. (CA3130B)*
- *Wide BW: 15 MHz typ. (unity-gain crossover)*
- *High SR: 10 V/μs typ. (unity-gain follower)*
- *High output current (I_O): 20 mA typ.*
- *High A_{OL}: 320,000 (110 dB) typ.*
- *Compensation with single external capacitor*

APPLICATIONS:

- *Ground-referenced single-supply amplifiers*
- *Fast sample-hold amplifiers*
- *Long-duration timers/monostables*
- *High-input-impedance comparators (ideal interface with digital COS/MOS)*
- *High-input-impedance wideband amplifiers*
- *Voltage followers (e.g., follower for single-supply D/A converter)*
- *Voltage regulators (permits control of output voltage down to zero volts)*
- *Peak detectors*
- *Single-supply full-wave precision rectifiers*
- *Photo-diode sensor amplifiers*

package (E suffix). All types operate over the full military-temperature range of -55°C to +125°C. The CA3130B is intended for applications requiring premium-grade specifications. The CA3130A offers superior input characteristics over those of the CA3130.

CA3130, CA3130A, CA3130B

ELECTRICAL CHARACTERISTICS at $T_A=25°C$, $V^+=15$ V, $V^-=0$ V (Unless otherwise specified)

CHARACTERISTIC	CA3130B (T,S)			CA3130A (T,S,E)			CA3130 (T,S,E)			Units
	Min.	Typ.	Max.	Min.	Typ.	Max.	Min.	Typ.	Max.	
Input Offset Voltage, $\|V_{IO}\|$, $V^\pm=\pm7.5$ V	–	0.8	2	–	2	5	–	8	15	mV
Input Offset Current, $\|I_{IO}\|$, $V^\pm=\pm7.5$ V	–	0.5	10	–	0.5	20	–	0.5	30	pA
Input Current, I_I $V^\pm=\pm7.5$ V	–	5	20	–	5	30	–	5	50	pA
Large-Signal Voltage Gain, A_{OL} $V_O=10$ V_{p-p}, $R_L=2$ kΩ	100 k	320 k	⌐	50 k	320 k	–	50 k	320 k	–	V/V
	100	110	–	94	110	–	94	110	–	dB
Common-Mode Rejection Ratio, CMRR	86	100	–	80	90	–	70	90	–	dB
Common-Mode Input-Voltage Range, V_{ICR}	0	-0.5 to 12	10	0	-0.5 to 12	10	0	-0.5 to 12	10	V
Power-Supply Rejection Ratio, $\Delta V_{IO}/\Delta V^\pm$ $V^\pm=\pm7.5$ V	–	32	100	–	32	150	–	32	320	µV/V
Maximum Output Voltage: At $R_L=2$ kΩ V_{OM}^+	12	13.3	–	12	13.3	–	12	13.3	–	V
V_{OM}^-	–	0.002	0.01	–	0.002	0.01	–	0.002	0.01	
At $R_L=\infty$ V_{OM}^+	14.99	15	–	14.99	15	–	14.99	15	–	
V_{OM}^-	–	0	0.01	–	0	0.01	–	0	0.01	
Maximum Output Current: I_{OM}^+ (Source) @ $V_O=0$ V	12	22	45	12	22	45	12	22	45	mA
I_{OM}^- (Sink) @ $V_O=15$ V	12	20	45	12	20	45	12	20	45	
Supply Current, I^+: $V_O=7.5$ V, $R_L=\infty$	–	10	15	–	10	15	–	10	15	mA
$V_O=0$ V, $R_L=\infty$	–	2	3	–	2	3	–	2	3	
Input Offset Voltage Temp. Drift, $\Delta V_{IO}/\Delta T^*$	–	5	–	–	10	–	–	10	–	µV/°C

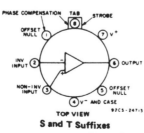

S and T Suffixes

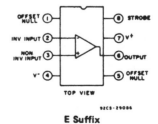

E Suffix

Fig.1 — Functional diagrams for the CA3130 series.

TYPICAL VALUES INTENDED ONLY FOR DESIGN GUIDANCE

CHARACTERISTIC	TEST CONDITIONS V^+ = +7.5 V V^- = −7.5 V T_A = 25°C (Unless Otherwise Specified)	CA3130B (T,S)	CA3130A (T,S,E)	CA3130 (T,S,E)	UNITS
Input Offset Voltage Adjustment Range	10 kΩ across Terms. 4 and 5 or 4 and 1	±22	±22	±22	mV
Input Resistance, R_I		1.5	1.5	1.5	TΩ
Input Capacitance, C_I	f = 1 MHz	4.3	4.3	4.3	pF
Equivalent Input Noise Voltage, e_n	BW = 0.2 MHz R_S = 1 MΩ*	23	23	23	μV
Unity Gain Crossover	C_C = 0	15	15	15	MHz
Frequency, f_T	C_C = 47 pF	4	4	4	
Slew Rate, SR: Open Loop	C_C = 0	30	30	30	V/μs
Closed Loop	C_C = 56 pF	10	10	10	
Transient Response: Rise Time, t_r	C_C = 56 pF C_L = 25 pF R_L = 2 kΩ (Voltage Follower)	0.09	0.09	0.09	μs
Overshoot		10	10	10	%
Settling Time (4 V_{p-p} Input to <0.1%)		1.2	1.2	1.2	μs

* Although a 1-MΩ source is used for this test, the equivalent input noise remains constant for values of R_S up to 10 MΩ.

CHARACTERISTIC	TEST CONDITIONS V^+ = 5 V V^- = 0 V T_A = 25°C (Unless Otherwise Specified)	CA3130B (T,S)	CA3130A (T,S,E)	CA3130 (T,S,E)	UNITS
Input Offset Voltage, V_{IO}		1	2	8	mV
Input Offset Current, I_{IO}		0.1	0.1	0.1	pA
Input Current, I_I		2	2	2	pA
Common-Mode Rejection Ratio, CMRR		100	90	80	dB
Large-Signal Voltage	V_O = 4 V_{p-p}	100 k	100 k	100 k	V/V
Gain, A_{OL}	R_L = 5 kΩ	100	100	100	dB
Common-Mode Input Voltage Range, V_{ICR}		0 to 2.8	0 to 2.8	0 to 2.8	V
Supply Current, I^+	V_O = 5 V, R_L = ∞	300	300	300	μA
	V_O = 2.5 V, R_L = ∞	500	500	500	
Power Supply Rejection Ratio, $\Delta V_{IO}/\Delta V^+$		200	200	200	μV/V

Wideband Operational Amplifier

Mini-DIP (E Suffix)
4-1817

8-Lead TO-5 With Dual-In-Line Formed Leads "DIL-CAN" (S Suffix)
H-1787

8-Lead TO-5 (T Suffix)
H-1528

Features:

- *High open-loop gain at video frequencies – 42 dB typ. at 1 MHz*
- *High unity-gain crossover frequency (f_T) – 38 MHz typ.*
- *Wide power bandwidth – $V_O = 18$ Vp-p typ. at 1.2 MHz*
- *High slew rate – 70 V/µs [typ.] in 20 dB amplifier 25 V/µs [typ.] in unity-gain amplifier*
- *Fast settling time – 0.6 µs typ.*
- *High output current – ±15 mA min.*
- *LM118, 748/LM101 pin compatibility*
- *Single capacitor compensation*
- *Offset null terminals*

Applications:

- *Video amplifiers*
- *Fast peak detectors*
- *Meter-driver amplifiers*
- *High-frequency feedback amplifiers*
- *Video pre-drivers*
- *Oscillators*
- *Multivibrators*
- *Voltage-controlled oscillator*
- *Fast comparators*

RCA-CA3100S, CA3100T is a large-signal wideband, high-speed operational amplifier which has a unity gain crossover frequency (f_T) of approximately 38 MHz and an open-loop, 3 dB corner frequency of approximately 110 kHz. It can operate at a total supply voltage of from 14 to 36 volts (±7 to ±18 volts when using split supplies) and can provide at least 18 Vp-p and 30 mA p-p at the output when operating from ±15 volt supplies. The CA3100 can be compensated with a single external capacitor and has dc offset adjust terminals for those applications requiring offset null. (See Fig. 15).

The CA3100 circuit contains both bipolar and P-MOS transistors on a single monolithic chip.

The CA3100 is supplied in either the standard 8-lead TO-5 package ("T" suffix), or in the 8-lead TO-5 dual-in-line formed-lead "DIL'CAN" package ("S" suffix).

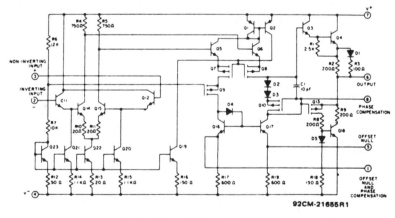

92CM-21655R1

Fig. 1 — Schematic diagram for CA3100.

CA3100 Types

ELECTRICAL CHARACTERISTICS, At $T_A = 25°C$:

CHARACTERISTICS	TEST CONDITIONS SUPPLY VOLTAGE (V$^+$,V$^-$)=15 V UNLESS OTHERWISE SPECIFIED	LIMITS MIN.	LIMITS TYP.	LIMITS MAX.	UNITS
STATIC					
Input Offset Voltage, V_{IO}	$V_O = 0 \pm 0.1$ V	–	± 1	± 5	mV
Input Bias Current, I_{IB}	$V_O = 0 \pm 1$ V	–	0.7	2	μA
Input Offset Current, I_{IO}		–	± 0.05	± 0.4	μA
Low-Frequency Open-Loop Voltage Gain, A_{OL} [•]	$V_O = \pm 1$ V Peak, F = 1 kHz	56	61	–	dB
Common-Mode Input Voltage Range, V_{ICR}	CMRR $>$ 76 dB	± 12	+14 / -13	–	V
Common-Mode Rejection Ratio, CMRR	V_I Common Mode = ± 12 V	76	90	–	dB
Maximum Output Voltage: Positive, V_{OM}^+	Differential Input Voltage = 0 ± 0.1 V	+9	+11	–	V
Negative, V_{OM}^-	R_L = 2 KΩ	–9	–11	–	
Maximum Output Current: Positive, I_{OM}^+	Differential Input Voltage = 0 ± 0.1 V	+15	+30	–	mA
Negative, I_{OM}^-	R_L = 250 Ω	–15	–30	–	
Supply Current, I^+	V_O = 0 ± 0.1 V, $R_L >$ 10 KΩ	-	8.5	10.5	mA
Power-Supply Rejection Ratio, PSRR	$\Delta V^+ = \pm 1$ V, $\Delta V^- = \pm 1$ V	60	70	–	dB
DYNAMIC					
Unity-Gain Crossover Frequency, f_T	C_C = 0, V_O = 0.3 V (P-P)	-	38	–	MHz
1-MHz Open-Loop Voltage Gain, A_{OL}	f = 1 MHz, C_C = 0, V_O = 10 V (P-P)	36	42	–	dB
Slew Rate, SR: 20-dB Amplifier	A_V = 10, C_C = 0, V_I = 1 V (Pulse)	50	70	–	V/μs
Follower Mode	A_V = 1, C_C = 10 pF, V_I = 10 V (Pulse)	–	25	–	
Power Bandwidth, PBW[▲]: 20-dB Amplifier	A_V = 10, C_C = 0, V_O = 18 V (P-P)	0.8	1.2	–	MHz
Follower Mode	A_V = 1, C_C = 10 pF, V_O = 18 V (P-P)	–	0.4	–	
Open-Loop Differential Input Impedance, Z_I	F = 1 MHz	–	30	–	KΩ
Open-Loop Output Impedance, Z_O	F = 1 MHz	–	110	–	Ω
Wideband Noise Voltage Referred to Input, e_N(Total)	BW = 1 MHz, R_S = 1 KΩ	–	8	–	μV RMS
Settling Time, t_s [To Within $\pm$ 50 mV of 9 V Output Swing]	R_L = 2 KΩ, C_L = 20 pF	–	0.6	–	μs

▲ Power Bandwidth = $\dfrac{\text{Slew Rate}}{\pi V_O \text{ (P-P)}}$ • Low-frequency dynamic characteristic

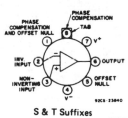

S & T Suffixes

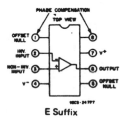

E Suffix

TERMINAL ASSIGNMENTS

MAXIMUM RATINGS, *Absolute-Maximum Values:*

Supply Voltage (between V+ and V⁻ terminals).	36	V
Differential Input Voltage	±12	V
Input Voltage to Ground●	±15	V
Offset Terminal to V⁻ Terminal Voltage.	±0.5	V
Output Current	50	mA●
Device Dissipation:		
Up to T_A = 55°C	630	mW
Above T_A = 55°C	6.67	mW/°C
Ambient Temperature Range:		
Operating:		
E Type	−40 to +85	°C
S and T Types	−55 to +125	°C
Storage	−65 to +150	°C
Lead Temperature (During Soldering):		
At distance 1/16 ± 1/32 inch (1.59 ± 0.79 mm) from case for 10 s max..	+265	°C

● If the supply voltage is less than ±15 volts, the maximum input voltage to ground is equal to the supply voltage.

● CA3100S, CA3100T does not contain circuitry to protect against short circuits in the output.

TYPICAL CHARACTERISTIC CURVES

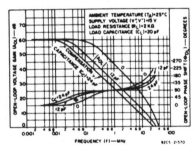

Fig. 2 – Open-loop gain, open-loop phase shift vs. frequency.

Fig. 3 – Open-loop gain vs. frequency and temperature.

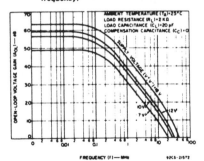

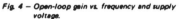

Fig. 4 – Open-loop gain vs. frequency and supply voltage.

Fig. 5 – Required compensation capacitance vs. closed-loop gain.

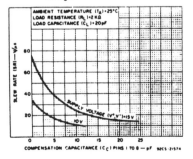

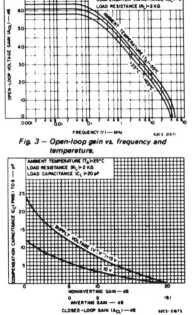

Fig. 6 – Slew rate vs. compensation capacitance.

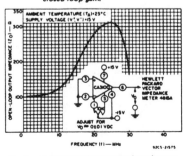

Fig. 7 – Typical open-loop output impedance vs. frequency.

TYPICAL CHARACTERISTIC CURVES (Cont'd)

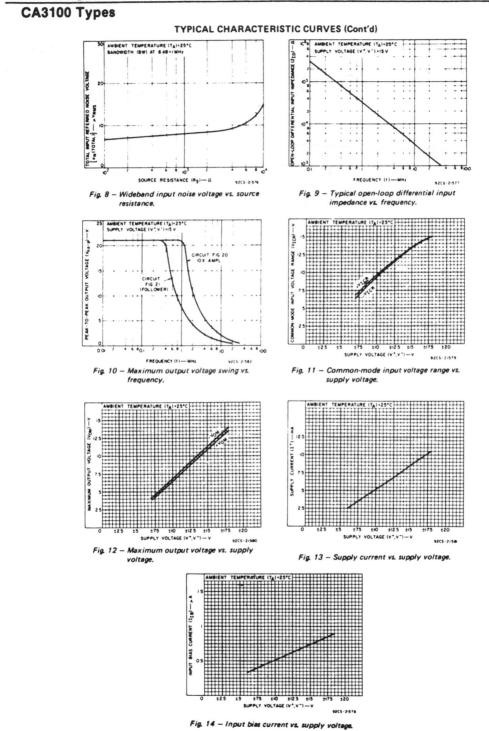

Fig. 8 – Wideband input noise voltage vs. source resistance.

Fig. 9 – Typical open-loop differential input impedance vs. frequency.

Fig. 10 – Maximum output voltage swing vs. frequency.

Fig. 11 – Common-mode input voltage range vs. supply voltage.

Fig. 12 – Maximum output voltage vs. supply voltage.

Fig. 13 – Supply current vs. supply voltage.

Fig. 14 – Input bias current vs. supply voltage.

Low-Power Wideband Amplifiers

12-Lead TO-5

H-1463

Features:
- Lower DC Power Drain:

P_T { CA3021 = 4 mW typ.
CA3022 = 12.5 mW typ.
CA3023 = 35 mW tp. } at Vcc = 6 V

- Excellent frequency response:

-3dB
BW { CA3021 = 2.4 MHz typ.
CA3022 = 7.5 MHz typ.
CA3023 = 16 MHz typ.

- High Voltage Gain:

A { CA3021 = 56 dB typ. at 0.5 MHz
CA3022 = 57 dB typ. at 2.5 MHz
CA3023 = 53 dB typ. at 5 MHz

- Wide AGC Range: 33 dB typ.
- Only one power supply (4.5 to 12 V) required
- Hermetically sealed 12-lead TO-5-style package
- Operation from −55° C to +125° C

Applications:
- Gain-controlled linear amplifiers
- AM/FM IF amplifiers
- Video amplifiers
- Limiters

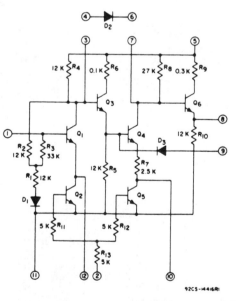

92CS-14416R1

Fig. 1 - Schematic diagram for CA3021, CA3022, and CA3023.

Linear Integrated Circuits

CA3021, CA3022, CA3023

ABSOLUTE-MAXIMUM RATINGS:

OPERATING-TEMPERATURE RANGE	-55°C to +125°C
STORAGE-TEMPERATURE RANGE	-65°C to +150°C
LEAD TEMPERATURE (During Soldering) At distance 1/16 ± 1/32 inch (1.59 ± 0.79mm) from case for 10 seconds max.	+265°C
DEVICE DISSIPATION, P_T	120 max. mW
INPUT-SIGNAL VOLTAGE	-3, +3 max. V
DC VOLTAGES AND CURRENTS	See Table Below

TERMINAL	VOLTAGE OR CURRENT LIMITS		CIRCUIT CONDITIONS	
	NEGATIVE	POSITIVE	TERMINAL	CONDITIONS
1	-3V	+3V	1	Connected to Voltage Source through 100Ω Resistor
			5	+12V
			10, 11, 12	Ground
2	-3V	+12V	5	+12V
			10, 11, 12	Ground
3	0V	+12V	5	+12V
			10, 11, 12	Ground
4	-12V +12V 10 max. mA		6, 11	Ground
5	0V	+18V	10, 11, 12	Ground
6	-12V +12V 10 max. mA		5, 11	Ground

TERMINAL	VOLTAGE OR CURRENT LIMITS		CIRCUIT CONDITIONS	
	NEGATIVE	POSITIVE	TERMINAL	CONDITIONS
7	0V	+12V	5	+12V
			10, 11, 12	Ground
8	20 max. mA		5	+12V
			10, 11, 12	Ground
9	-0.5V	+3V	5	+12V
			10, 11, 12	Ground
10	0V	+4V	2,5	+12V
			11	Ground
11	-6V	+12V	2	Ground
			5	+12V
12	0V	+4V	2,5	+12V
			11	Ground

213

ELECTRICAL CHARACTERISTICS, at $T_A = 25^\circ C$, $V_{CC} = +6V$, unless otherwise specified

CHARACTERISTIC	SYMBOL	TEST CONDITIONS			LIMITS									UNITS	TYPICAL CHARACTERISTIC CURVE
		TEST SETUP AND PROCEDURE	FEEDBACK RESISTANCE (R_β) BETWEEN TERMINALS 3 AND 7	FREQUENCY f	CA3021 (TA5219)			CA3022 (TA5236)			CA3023 (TA5218)				
		Fig.	Ω	MHz	Min.	Typ.	Max.	Min.	Typ.	Max.	Min.	Typ.	Max.	Units	Fig.
Device Dissipation	P_T	2	∞	–	1	4	8	–	–	–	–	–	–	mW	3a,d
			∞	–	–	–	–	5	12.5	24	–	–	–	mW	3b,d
			∞	–	–	–	–	–	–	–	24	35	48	mW	3c,d
Quiescent Output Voltage	V_O	2	39k	–	–	2.2	–	–	–	–	–	–	–	V	–
			10k	–	–	–	–	–	1.9	–	–	–	–	V	
			4.7k	–	–	–	–	–	–	–	–	1.3	–	V	
AGC Source Current	I_{AGC}	4	$V_{AGC} = +6V$	–	–	0.8	–	–	0.8	–	–	0.8	–	mA	–
Voltage Gain	A	5	560k	0.5	50	56	–	–	–	–	–	–	–	dB	6a
			39k	0.8	40	46	–	–	–	–	–	–	–	dB	6a,d
			39k	2.5	–	–	–	50	57	–	–	–	–	dB	6b
			10k	3	–	–	–	40	44	–	–	–	–	dB	6b,d
			18k	5	–	–	–	–	–	–	50	53	–	dB	6c
			4.7k	10	–	–	–	–	–	–	40	44	–	dB	6c,d
Bandwidth at -3 dB Point	BW	5	39k	–	0.8	2.4	–	–	–	–	–	–	–	MHz	6a
			10k	–	–	–	–	3	7.5	–	–	–	–	MHz	6b
			4.7k	–	–	–	–	–	–	–	10	16	–	MHz	6c
Input-Impedance Components — Input Resistance	R_{IN}	7	39k	1	–	4000	–	–	–	–	–	–	–	Ω	–
			10k	5	–	–	–	–	1300	–	–	–	–	Ω	
			4.7k	10	–	–	–	–	–	–	–	300	–	Ω	
Input-Impedance Components — Input Capacitance	C_{IN}	7	39k	1	–	11	–	–	–	–	–	–	–	pF	–
			10k	5	–	–	–	–	18	–	–	–	–	pF	
			4.7k	10	–	–	–	–	–	–	–	13	–	pF	
Output Resistance	R_{OUT}	8	39k	1	–	300	–	–	–	–	–	–	–	Ω	–
			10k	5	–	–	–	–	120	–	–	–	–	Ω	
			4.7k	10	–	–	–	–	–	–	–	100	–	Ω	
Noise Figure	NF	9	39k	1	–	4.2	8.5	–	–	–	–	–	–	dB	–
			10k	1	–	–	–	–	4.4	8.5	–	–	–	dB	
			4.7k	1	–	–	–	–	–	–	–	6.5	8.5	dB	
AGC Range	AGC	10	–	1	–	33	–	–	–	–	–	–	–	dB	–
			–	5	–	–	–	–	33	–	–	–	–	dB	
			–	10	–	–	–	–	–	–	–	33	–	dB	
Maximum Output Voltage (RMS Value)	v_{out}	5	39k	1	–	0.6	–	–	–	–	–	–	–	V(rms)	–
			10k	5	–	–	–	–	0.7	–	–	–	–	V(rms)	
			4.7k	10	–	–	–	–	–	–	–	0.5	–	V(rms)	

CA3021, CA3022, CA3023

TEST SETUP FOR MEASUREMENT OF DEVICE DISSIPATION
AND QUIESCENT OUTPUT VOLTAGE

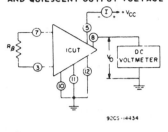

$$P_T = V_{CC} (I)$$

Fig.2

DEVICE DISSIPATION VS DC SUPPLY VOLTAGE
FOR CA3021

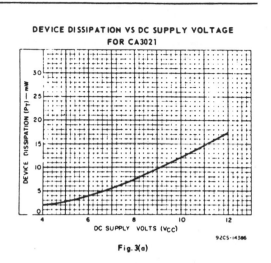

Fig.3(a)

DEVICE DISSIPATION VS DC SUPPLY VOLTAGE
FOR CA3022

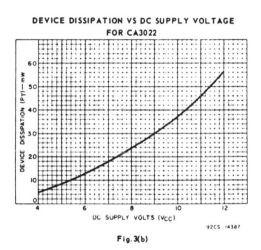

Fig.3(b)

DEVICE DISSIPATION VS DC SUPPLY VOLTAGE
FOR CA3023

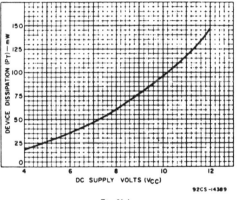

Fig.3(c)

DEVICE DISSIPATION VS TEMPERATURE FOR
CA3021, CA3022, AND CA3023

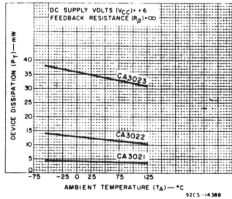

Fig.3(d)

TEST SETUP FOR MEASUREMENT OF AGC
SOURCE CURRENT

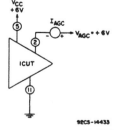

I_{AGC} IS THE CURRENT FLOWING INTO TERMINAL 2.

Fig.4

215

TEST SETUP FOR MEASUREMENTS OF VOLTAGE-GAIN, -3dB BANDWIDTH, AND MAXIMUM OUTPUT VOLTAGE

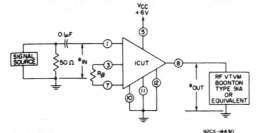

PROCEDURES

Voltage Gain:

(a) Set e_{in} = 0.5 mV at frequency specified, read e_{out} Voltage Gain

$(A) = 20 \, Log_{10} \dfrac{e_{out}}{e_{in}}$

Bandwidth:

(a) Set e_{out} to a convenient reference voltage at f = 100 kHz and record corresponding value of e_{in}.

(b) Increase the frequency, keeping e_{in} constant until e_{out} drops 3-dB. Record Bandwidth.

Fig.5

VOLTAGE GAIN VS FREQUENCY FOR CA3021

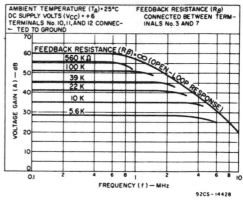

Fig.6(a)

VOLTAGE GAIN VS FREQUENCY FOR CA3022

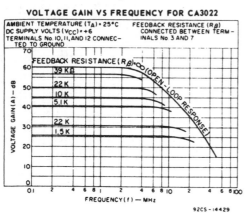

Fig.6(b)

VOLTAGE GAIN VS FREQUENCY FOR CA3023

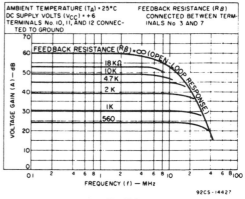

Fig.6(c)

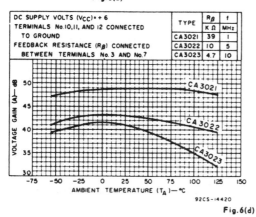

TYPE	R_β KΩ	f MHz
CA3021	39	1
CA3022	10	5
CA3023	4.7	10

Fig.6(d)

VOLTAGE GAIN VS TEMPERATURE FOR CA3021, CA3022, AND CA3023

APPENDIX B
Passive Component Values and Identification

Appendix B gives Tables, Charts, and Drawings for component values and tolerances and identification of these values and tolerances. Included in the components are resistors, capacitors, and diodes.

1. Tolerance and component values of carbon-composition resistors
2. Resistor color code
3. Resistor types and ranges
4. Capacitor color code
5. Capacitor ranges on the basis of dielectric material
6. Semiconductor diode identification

TABLE 1 Tolerance and component values of carbon-composition resistors

20% tolerance	10% tolerance	5% tolerance
10	10	10
	12	12
15	15	15
		16
	18	18
		20
22	22	22
		24
	27	27
		30
33	33	33
		36
	39	39
		43
47	47	47
		51
	56	56
		62
68	68	68
		75
	82	82
		91
100	100	100

Harry E. Thomas, *Handbook for Electronic Engineers and Technicians*, copyright 1965. Reprinted by permission of Prentice Hall, Inc.

Table 2 Resistor color code

A B C D

$R = AB \times 10^C, D$

A, B, C digits

Black	0	Green	5
Brown	1	Blue	6
Red	2	Violet	7
Orange	3	Gray	8
Yellow	4	White	9
		Gold (C only)	-1
		Silver (C only)	-2

Example

| A | Red | C | Orange |
| B | Violet | D | Gold |

$R = 27 \times 10^3, 5\%$
$= 27\ k\Omega, 5\%$

Tolerance (D) code

No band	20%
Silver	10%
Gold	5%

Bruce D. Wedlock and James K. Roberge, *Electronic Components and Measurements*, copyright © 1978. Reprinted by permission of Prentice Hall, Inc.

TABLE 3 Resistor types and ranges

Resistor Type		General-Purpose	Semi-precision	Precision	Ultra-precision	Power Ratings	Resistance Range (Standard)
	Carbon composition	+				1/8 to 2W	
	Cermet		+			1/8 to 2 W	
Film	Carbon-film		+			1/2 to 2 W	
	Metal-oxide-film			+	+	1/10 to 2 W	
	Metal-film			+	+	1/20 to 2 W	
	Phenolic	+				1 to 2 W	
	Ceramic-shell	+				2 to 22 W	
	Flameproof	+				1 to 10 W	
	Vitreous-enamel	+				1 to 11 W	
Small Axial-Lead Wirewound	Silicone-ceramic		+	+		1 to 15 W	
	Silicone-coated	+	+			0.25 to 15 W	
	Bobbin			+		0.10 to 1 W	
	Heatsinked		+	+		7.50 to 100 W	
Power Wirewound	Tubular	+	+			3 to 225 W	
	Thin	+	+			10 to 55 W	
	Tubular-adjustable	+	+			12 to 225 W	
	High-voltage/high-resistance	+	+			1 to 90 W	

Resistance Range (Standard) scale columns: .1 Ω, 1 Ω, 10 Ω, 100 Ω, 1 kΩ, 10 kΩ, 100 kΩ, 1 MΩ, 10 MΩ, 100 MΩ, 1000 MΩ

Thomas H. Jones, *Electronic Components Handbook*, copyright © 1978. Reprinted by permission of Prentice Hall, Inc.

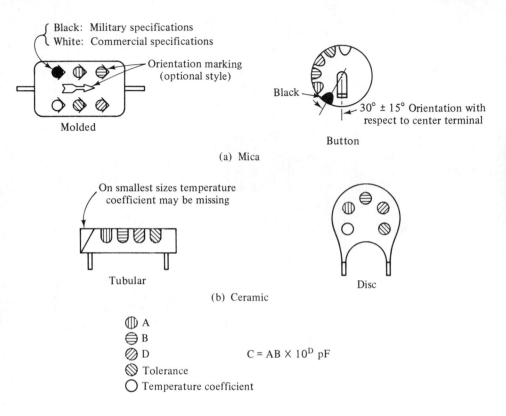

{ Black: Military specifications
{ White: Commercial specifications

Orientation marking
(optional style)

Molded

Black —

30° ± 15° Orientation with
respect to center terminal

Button

(a) Mica

On smallest sizes temperature
coefficient may be missing

Tubular

Disc

(b) Ceramic

A
B
D $C = AB \times 10^{D}$ pF
Tolerance
Temperature coefficient

TABLE 4 Capacitor Color Codes

						CERAMIC		
		MICA						
						Tolerance		Temp. Coeff.
Color	A,B	D	Tolerance %	Temp. Coeff. ppm/°C	D	% (>10 pF)	pF (>10 pF)	ppm/°C
Black	0	0	±20 M		0	±20 M	±2.0 G	0
Brown	1	1	±1 F	±500	1	±1 F	±0.1 B	−33
Red	2	2	±2 G	±200	2	±2 G		−75
Orange	3	3		±100	3	±3 H		−150
Yellow	4	4		−20 to +100	4			−220
Green	5		±5 J	0 to +70		±5 J	±0.5 D	−330
Blue	6							−470
Violet	7							−750
Gray	8				−2		±0.25 C	−1500 to +150
White	9				−1	±10 K	±1.0 F	−750 to +100
Gold		−1	±0.5 E					
Silver		−2	±10 K					

Bruce D. Wedlock and James K. Roberge, *Electronic Components and Measurements*, copyright © 1978.
Reprinted by permission of Prentice Hall, Inc.

TABLE 5 Capacitor ranges on the basis of dielectric material

Capacitance

	1 F	100,000 µF	10,000 µF	1,000 µF	100 µF	10 µF	1 µF	0.1 µF	0.01 µF	0.001	100 pF	10 pF	1 pF
Paper and plastic													
Paper													
Metalized paper													
Polyester													
Metalized polyester													
Polyester and paper													
Polystyrene													
Polycarbonate													
Polypropylene													
Metalized polypropylene													
Teflon®													
Mica and glass													
Mica													
Transmitting mica													
Glass													
Glass-K													
Ceramic													
Class I disc (T.C.)													
Class II disc													
Semiconductor disc													
Class I tubular (T.C.)													
Class II tubular													
Monolithic (T.C.)													
Monolithic plate													
Monolithic tubular													
Monolithic molded													
Aluminum													
Twist mount													
Tubular													
Miniature tubular													
Computer grade													
Computer grade, axial													
Tantalum													
Plain foil													
Etched foil													
Wet slug													
Dry electrolyte													
Oil-filled													
DC													
AC													
AC-DC													
Energy discharge													

Thomas H. Jones, *Electronic Components Handbook,* copyright © 1978. Reprinted by permission of Prentice Hall, Inc.

Table 6 Semiconductor diode identification

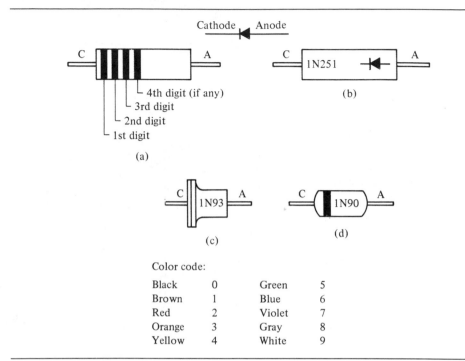

Cathode   Anode

C   A

4th digit (if any)
3rd digit
2nd digit
1st digit

(a)

C   1N251   A

(b)

C   1N93   A

(c)

C   1N90   A

(d)

Color code:

Black	0	Green	5
Brown	1	Blue	6
Red	2	Violet	7
Orange	3	Gray	8
Yellow	4	White	9

(Reprinted by permission of Prentice Hall, Inc.)

Index